ENCYCLOPAEDIA OF OUTLINE PHYSICAL GEOGRAPHY

ENCYCLOPAEDIA OF OUTLINE PHYSICAL GEOGRAPHY

R.V. Prajapati

CYBER TECH PUBLICATIONS
4264/3, Ansari Road, Daryaganj, New Delhi-110002 (India)
Ph.: 011-23244078, 011-43559448 Fax: 011-23280028
E-Mail: cyberpublicationsdelhi@yahoo.com
Website: www.cybertechpublications.com

ENCYCLOPAEDIA OF OUTLINE PHYSICAL GEOGRAPHY

R.V. Prajapati

First Edition 2010

Published by :

G.S. Rawat for **Cyber Tech Publications**

4264/3, Ansari Road, Daryaganj, New Delhi-110002 (India)

Ph.: 011-23244078, 011-43559448 Fax: 011-23280028

E-Mail: cyberpublicationsdelhi@yahoo.com

Website: www.cybertechpublications.com

Printed at: Asian Offset Printers, Delhi-53

Preface

Physical Geography is a sub field of geography that focuses on the systematic study of patterns and processes within the hydrosphere biosphere atmosphere and lithosphere. It aims to understand the physical layout of the earth its weather and global flora and fauna patterns. Many areas of physical geography make use of geology particularly in the study of weathering and erosion. The geology of other planets is discussed at Geological features of the solar system. Physical geography as a scientific discipline is usually contrasted with and complemented by its sister science human geography.

Over geological time the earth has changed physically. A product of these changes was the creation of conditions in which life became possible. Once established, the burgeoning life-forms which the earth supported became a part of the substantive matrix of earth evolution. Some of the latest-developed of these opportunistic life-forms were small, bipedal apes. From the forests and savannahs of Africa, some of the hominid ancestors of these apes evolved into people. As people, and like the other life-forms with which they shared the planet, they responded in different ways to the environment of which they had become a part. And out of these responses, empowered by the development of language and bonded by the crystallization of ritual into religion, their influence upon it began to enlarge. Human groups became culturally differentiated: some became sedentary as the realization of agricultural systems and incipient urbanization signalled the possibilities of societal enlargement and the extension of socio-political ambition through the control of other people and of earth space. It was some of these groups that were to be primarily responsible for 'The Great Transformation'.

'The Great Transformation' is a topical theme in geography. The argument runs that the ecosphere has suffered changes of such magnitude and variety at the hands of the human species that its fundamental appearance and nature have been, or are about to be, radically altered. It would be wrong to belittle this view; indeed, much evidence in its favour will emerge later in the encyclopedia. None the less, it is right from the full perspective of geological time to note that the Great Transformation is but one of the changes, though possibly one of the most radical, that have occurred in the ecosphere during its history. All living beings, and not just the human species, actively change their environment. Since life first evolved, the living and non-living worlds have interacted. The course of biological evolution has not been steered solely by environmental change: life itself changes its environment and to a large extent determines the conditions in the air, sea, and land surface where it thrives.

The distinctiveness of the geographical discipline has always lain in the extent to which it accommodates a concern for the physical as well as for the human environment and in its early vision of an integrated and holistic world: a habitat; a place of organic life. In practice, and in relatively recent times, this view has been severely strained by the widening gap between physical and human geography, by the tensions created between competing explanatory paradigms—especially in human geography—and by a multiplicity of steadily narrowing specialisms. Superficially, the outcome may seem to threaten the academic integrity and coherence of the discipline. Yet this coherence is arguably more necessary now than at any time in the past, as habitat pressures strain the weakening seams of a stretched earth. This encyclopedia, therefore, offers a view of geography from an earlier, wider, and sometimes bolder perspective by bringing together a wide variety of contributions from scholars of both the physical and human environments, and by focusing their expertise on the large theme of the evolution of the earth as a habitat. In the process, they illustrate not only the essential character of the geographical tradition but the centrality of its holistic vision to the most fundamental problems of habitat change and survival.

Contents

1

INTRODUCTION

Some of the first truly geographical studies occurred more than four thousand years ago. The main purpose of these early investigations was to map features and places observed as explorers traveled to new lands. At this time, Chinese, Egyptian, and Phoenician civilizations were beginning to explore the places and spaces within and outside their homelands. The earliest evidence of such explorations comes from the archaeological discovery of a Babylonian clay tablet map that dates back to 2300 BC.

The early Greeks were the first civilization to practice a form of geography that was more than mere map making or cartography. Greek philosophers and scientist were also interested in learning about spatial nature of human and physical features found on the Earth. One of the first Greek geographers was Herodotus (circa 484 - 425 BC). Herodotus wrote a number of volumes that described the human and physical geography of the various regions of the Persian Empire.

The ancient Greeks were also interested in the form, size, and geometry of the Earth. Aristotle (circa 384 - 322 BC) hypothesized and scientifically demonstrated that the Earth had a spherical shape. Evidence for this idea came from observations of lunar eclipses. Lunar eclipses occur when the Earth casts its circular shadow on to the moon's surface. The first individual to accurately calculate the circumference of the Earth was the Greek geographer Eratosthenes (circa 276 - 194 BC). Eratosthenes calculated the equatorial circumference to be 40,233 kilometers using simple geometric relationships. This primitive calculation was unusually accurate. Measurements of the Earth using modern satellite technology have computed the circumference to be 40,072 kilometers.

Most of the Greek accomplishments in geography were passed on to the Romans. Roman military commanders and administrators used this information to guide the expansion of their Empire. The Romans also made several important additions to geographical knowledge. Strabo (circa 64 BC

- 20 AD) wrote a 17 volume series called "Geographia". Strabo claimed to have traveled widely and recorded what he had seen and experienced from a geographical perspective. In his series of books, Strabo describes the cultural geographies of the various societies of people found from Britain to as far east as India, and south to Ethiopia and as far north as Iceland. Strabo also suggested a definition of geography that is quite complementary to the way many human geographers define their discipline today. This definition suggests that the aim of geography was to "describe the known parts of the inhabited world ... to write the assessment of the countries of the world [and] to treat the differences between countries".

During the second century AD, Ptolemy (circa 100 - 178 AD) made a number of important contributions to geography. Ptolemy's publication Geographike hyphegesis or "Guide to Geography" compiled and summarize much of the Greek and Roman geographic information accumulated at that time. Some of his other important contributions include the creation of three different methods for projecting the Earth's surface on a map, the calculation of coordinate locations for some eight thousand places on the Earth, and development of the concepts of geographical latitude and longitude.

Figure: This early map of the world was constructed using map making techniques developed by Ptolemy. Note that the map is organized with crisscrossing lines of latitude and longitude.

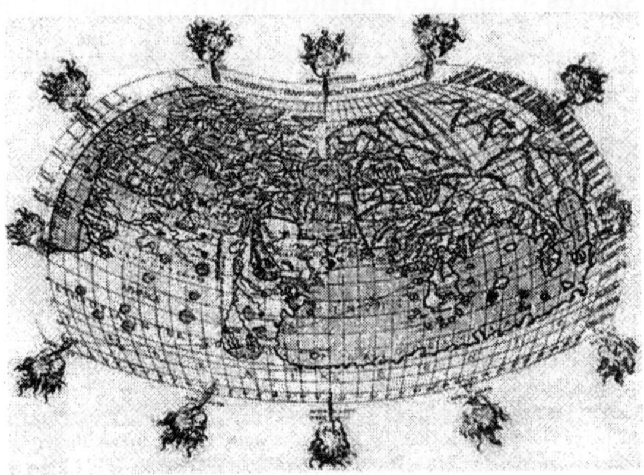

Little academic progress in geography occurred after the Roman period. For the most part, the Middle Ages (5th to 13th centuries AD) were a time of intellectual stagnation. In Europe, the Vikings of Scandinavia were the only group of people carrying out active exploration of new lands. In the Middle East, Arab academics began translating the works of Greek and Roman geographers starting in the 8th century and began exploring southwestern Asia and Africa.

Some of the important intellectuals in Arab geography were Al-Idrisi, Ibn Battutah, and Ibn Khaldun. Al-Idrisi is best known for his skill at making maps and for his work of descriptive geography Kitab nuzhat al-mushtaq fi ikhtiraq al-afaq or "The Pleasure Excursion of One Who Is

Eager to Traverse the Regions of the World". Ibn Battutah and Ibn Khaldun are well known for writing about their extensive travels of North Africa and the Middle East.

During the Renaissance (1400 to 1600 AD) numerous journeys of geographical exploration were commissioned by a variety of nation states in Europe. Most of these voyages were financed because of the potential commercial returns from resource exploitation. The voyages also provided an opportunity for scientific investigation and discovery. These voyages also added many significant contributions to geographic knowledge. Important explorers of this period include Christopher Columbus, Vasco da Gama, Ferdinand Magellan, Jacques Cartier, Sir Martin Frobisher, Sir Francis Drake, John and Sebastian Cabot, and John Davis. Also during the Renaissance, Martin Behaim created a spherical globe depicting the Earth in its true three-dimensional form in 1492. Behaim's invention was a significant advance over two-dimensional maps because it created a more realistic depiction of the Earth's shape and surface configuration.

Figure: This map was constructed by Oliva in 1560. It describes the known world at this time and suggests that North America is part of Asia. Further exploration of the world would soon reject this idea.

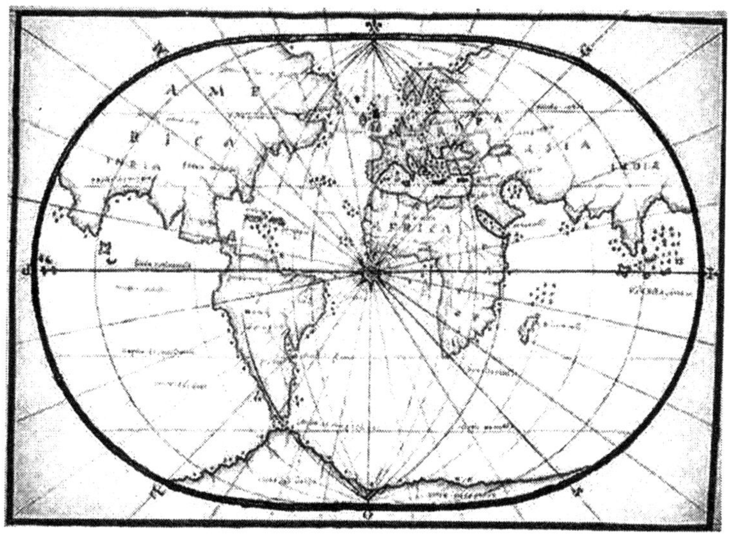

In the 17th century, Bernhardus Varenius (1622-1650) published an important geographic reference titled Geographia generalis (General Geography: 1650). In this volume, Varenius used direct observations and primary measurements to present some new ideas concerning geographic knowledge. This work continued to be a standard geographic reference for about a 100 years. Varenius also suggested that the discipline of geography could be subdivided into three distinct branches. The first branch examines the form and dimensions of the Earth. The second sub-discipline deals with tides, climatic variations over time and space, and other variables that are influenced by the cyclical movements of the Sun and moon. Together these two branches form the early beginning of what we collectively now call physical geography. The last branch of geography examined distinct regions on the Earth using comparative cultural studies. Today, this area of knowledge is called cultural geography.

During the 18th century, the German philosopher Immanuel Kant (1724-1804) proposed that human knowledge could be organized in three different ways. One way of organizing knowledge was to classify its facts according to the type of objects studied. Accordingly, zoology studies animals, botany examines plants, and geology involves the investigation of rocks. The second way one can study things is according to a temporal dimension. This field of knowledge is of course called history. The last method of organizing knowledge involves understanding facts relative to spatial relationships. This field of knowledge is commonly known as geography. Kant also divided geography into a number of sub-disciplines. He recognized the following six branches: Physical, mathematical, moral, political, commercial, and theological geography.

Geographic knowledge saw strong growth in Europe and the United States in the 1800s. This period also saw the emergence of a number of societies interested in geographic issues. In Germany, Alexander von Humboldt, Carl Ritter, and Fredrich Ratzel made substantial contributions to human and physical geography. Humboldt's publication Kosmos (1844) examines the geology and physical geography of the Earth. This work is considered by many academics to be a milestone contribution to geographic scholarship. Late in the 19th Century, Ratzel theorized that the distribution and culture of the Earth's various human populations was strongly influenced by the natural environment. The French geographer Paul Vidal de la Blanche opposed this revolutionary idea.

Instead, he suggested that human beings were a dominant force shaping the form of the environment. The idea that humans were modifying the physical environment was also prevalent in the United States. In 1847, George Perkins Marsh gave an address to the Agricultural Society of Rutland County, Vermont. The subject of this speech was that human activity was having a destructive impact on land, especially through deforestation and land conversion. This speech also became the foundation for his book Man and Nature or The Earth as Modified by Human Action, first published in 1864. In this publication, Marsh warned of the ecological consequences of the continued development of the American frontier.

During the first 50 years of the 1900s, many academics in the field of geography extended the various ideas presented in the previous century to studies of small regions all over the world. Most of these studies used descriptive field methods to test research questions. Starting in about 1950, geographic research experienced a shift in methodology. Geographers began adopting a more scientific approach that relied on quantitative techniques. The quantitative revolution was also associated with a change in the way in which geographers studied the Earth and its phenomena. Researchers now began investigating process rather than mere description of the event of interest. Today, the quantitative approach is becoming even more prevalent due to advances in computer and software technologies.

In 1964, William Pattison published an article in the Journal of Geography (1964, 63: 211-216) that suggested that modern Geography was now composed of the following four academic traditions:

- Spatial Tradition - the investigation of the phenomena of geography from a strictly spatial perspective.
- Area Studies Tradition - the geographical study of an area on the Earth at either the local, regional, or global scale.
- Human-Land Tradition - the geographical study of human interactions with the environment.
- Earth Science Tradition - the study of natural phenomena from a spatial perspective. This tradition is best described as theoretical physical geography.

Today, the academic traditions described by Pattison are still dominant fields of geographical investigation. However, the frequency and magnitude of human mediated environmental problems has been on a steady increase since the publication of this notion. These increases are the result of a growing human population and the consequent increase in the consumption of natural resources. As a result, an increasing number of researchers in geography are studying how humans modify the environment. A significant number of these projects also develop strategies to reduce the negative impact of human activities on nature. Some of the dominant themes in these studies include: environmental degradation of the hydrosphere, atmosphere, lithosphere, and biosphere; resource use issues; natural hazards; environmental impact assessment; and the effect of urbanization and land-use change on natural environments. Considering all of the statements presented concerning the history and development of geography, we are now ready to formulate a somewhat coherent definition. This definition suggests that geography, in its simplest form, is the field of knowledge that is concerned with how phenomena are spatially organized. Physical geography attempts to determine why natural phenomena have particular spatial patterns and orientation. This online textbook will focus primarily on the Earth Science Tradition. Some of the information that is covered in this textbook also deals with the alterations of the environment because of human interaction. These pieces of information belong in the Human-Land Tradition of geography.

ELEMENTS OF GEOGRAPHY

In the previous section, we discovered that geography consists of at least two different sub-fields of knowledge with similar methodology: Physical geography and human geography. The following table also helps to make the differences between these two types of geography more apparent. This table describes some of the phenomena or elements studied by each of these sub-fields of knowledge. Knowing what kinds of things are studied by geographers provides us with a better understanding of the differences between physical and human geography.

Table: Some of the phenomena studied in physical and human geography.

Physical Geography	Human Geography
Rocks and Minerals	Population
Landforms	Settlements
Soils	Economic Activities
Animals	Transportation
Plants	Recreational Activities
Water	Religion
Atmosphere	Political Systems
Rivers and Other Water Bodies	Social Traditions
Environment	Human Migration
Climate and Weather	Agricultural Systems
Oceans	Urban Systems

Figure: Major subdisciplines of physical and human geography.

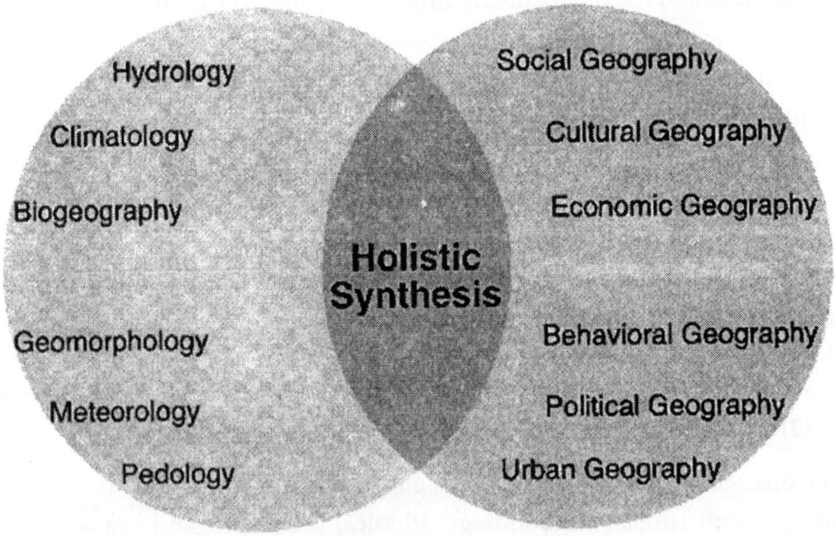

Geography is also a discipline that integrates a wide variety of subject matter. Almost any area of human knowledge can be examined from a spatial perspective. Figure 1b-1 describes some of the main subdisciplines within human and physical geography. Physical geography's primary subdisplines study the Earth's atmosphere (meteorology and climatology), animal and plant life (biogeography), physical landscape (geomorphology), soils (pedology), and waters (hydrology). Some of the dominant areas of study in human geography include: human society and culture (social and cultural geography), behavior (behavioral geography), economics (economic geography), politics (political geography), and urban systems (urban geography).

The graphic model in above Figure indicates that the study of geography can also involve a holistic synthesis. Holistic synthesis connects knowledge from a variety of academic fields in both human and physical geography. For example, the study of the enhancement of the Earth's greenhouse effect and the resulting global warming requires a multidisciplinary approach for complete understanding. The fields of climatology and meteorology are required to understand the physical effects of adding addition greenhouse gases to the atmosphere's radiation balance. The field of economic geography provides information on how various forms of human economic activity contribute to the emission of greenhouse gases through fossil fuel burning and land-use change. Combining the knowledge of both of these academic areas gives us a more comprehensive understanding of why this serious environmental problem occurs.

The holistic nature of geography is both a strength and a weakness. Geography's strength comes from its ability to connect functional interrelationships that are not normally noticed in narrowly defined fields of knowledge. The most obvious weakness associated with the geographical approach is related to the fact that holistic understanding is often too simple and misses important details of cause and effect.

Scope of Physical Geography

We have now learned that physical geography examines and investigates natural phenomena spatially. In the previous section, we identified some of the key elements studied by physical geographers. Combining these two items, we can now suggest that physical geography studies the spatial patterns of weather and climate, soils, vegetation, animals, water in all its forms, and landforms. Physical geography also examines the interrelationships of these phenomena to human activities. This sub-field of geography is academically known as the Human-Land Tradition. This area of geography has seen very keen interest and growth in the last few decades because of the acceleration of human induced environmental degradation. Thus, physical geography's scope is much broader than the simple spatial study of nature. It also involves the investigation of how humans are influencing nature.

Academics studying physical geography and other related earth sciences are rarely generalists. Most are in fact highly specialized in their fields of knowledge and tend to focus themselves in one of the following well defined areas of understanding in physical geography:

- Geomorphology - studies the various landforms on the Earth's surface.
- Pedology - is concerned with the study of soils.

- Biogeography - is the science that investigates the spatial relationships of plants and animals.
- Hydrology - is interested in the study of water in all its forms.
- Meteorology - studies the circulation of the atmosphere over short time spans.
- Climatology - studies the effects of weather on life and examines the circulation of the atmosphere over longer time spans.

The above fields of knowledge generally have a primary role in introductory textbooks dealing with physical geography. Introductory physical geography textbooks can also contain information from other related disciplines including:

- Geology - studies the form of the Earth's surface and subsurface, and the processes that create and modify it.
- Ecology - the scientific study of the interactions between organisms and their environment.
- Oceanography - the science that examines the biology, chemistry, physics, and geology of oceans.
- Cartography - the technique of making maps.
- Astronomy - the science that examines celestial bodies and the cosmos.

Geography as an Environmental Science

Webster's 9th Collegiate Dictionary defines Environment "... as the complex of physical, chemical, and biotic factors (such as climate, soil, and living things) that act upon an organism or an ecological community and ultimately determines its form and survival".

Both human and physical geography provide an important intellectual background for studying the environment. Many environmental studies/science programs offered by Universities and Colleges around the world rely on the information found in various geography courses to help educate their students about the state of the environment.

History of Physical Geography

The nature of understanding in physical geography has changed over time. When investigating this change it becomes apparent that certain universal ideas or forces had very important ramifications to the academic study of physical geography. During the period from 1850 to 1950, there seems to be four main ideas that had a strong influence on the discipline:

1. *Uniformitarianism* - this theory rejected the idea that catastrophic forces were responsible for the current conditions on the Earth. It suggested instead that continuing uniformity of existing processes were responsible for the present and past conditions of this planet.

2. *Evolution* - Charles Darwin's Origin of Species (1859) suggested that natural selection determined which individuals would pass on their genetic traits to future generations. As a result of this theory, evolutionary explanations for a variety of natural phenomena were postulated by scientists. The theories of uniformitarianism and evolution arose from a fundamental change in the way humans explained the universe and nature. During the 16th, 17th, and 18th centuries scholars began refuting belief or myth based explanations of the cosmos, and instead used science to help explain the mysteries of nature. Belief based explanations of the cosmos are made consistent with a larger framework of knowledge that focuses on some myth. However, theories based on science questioned the accuracy of these beliefs.
3. *Exploration and Survey* - much of the world had not been explored before 1900. Thus, during this period all of the fields of physical geography were actively involved with basic data collection. This data collection included activities like determining the elevation of land surfaces, classification and description of landforms, the measurement of the volume of flow of rivers, measurement of various phenomena associated to weather and climate, and the classification of soils, organisms, biological communities and ecosystems.
4. *Conservation* - beginning in the 1850s a concern for the environment began to develop as a result of the human development of once natural areas in the United States and Europe. One of the earliest statements of these ideas came from George Perkins Marsh (1864) in his book "Man in Nature" or "Physical Geography as Modified by Human Action". This book is often cited by scholars as the first significant academic contribution to conservation and environmentalism.

After 1950, the following two forces largely determined the nature of physical geography:

1. *The Quantitative Revolution* - measurement became the central focus of research in Physical Geography. It was used primarily for hypothesis testing. With measurement came mapping, models, statistics, mathematics, and hypothesis testing. The quantitative revolution was also associated with a change in the way in which physical geographers studied the Earth and its phenomena. Researchers now began investigating process rather than mere description of the environment.
2. *The study of Human/Land Relationships* - the influence of human activity on the environment was becoming very apparent after 1950. As a result, many researchers in physical geography began studying the influence of humans on the environment. Some of the dominant themes in these studies included: environmental degradation and resource use; natural hazards and impact assessment; and the effect of urbanization and land-use change on natural environments.

FUTURE OF PHYSICAL GEOGRAPHY

The following list describes some of the important future trends in physical geography research:

- Continued development of applied physical geography for the analysis and correction of human-induced environmental problems. A student of applied physical geography uses theoretical information from the field of physical geography to manage and solve problems related to natural phenomena found in the real world.
- Remote Sensing - Advances in technology have caused the development of many new instruments for the monitoring of the Earth's resources and environment from airborne and space platforms. The most familiar use of remote sensing technology is to monitor the Earth's weather for forecasting. See the following links for examples of remotely sensed images:
    - o Google Maps
    - o Google Earth
    - o France - SPOT Image
    - o USA - Landsat
    - o USA - MODIS
    - o USA - ASTER
    - o Canada - Radarsat
    - o GOES
    - o USA EROS Data Center
- Geographic Information Systems - A geographic information system (GIS) merges information in a computer database with spatial coordinates on a digital map. Geographic information dystems are becoming increasingly more important for the management of resources.

Physical geography is done as much by boots and compass as by mental activity. Philosophy in physical geography is an active process open to change as the subject is practised. This book taps into philosophy in action in physical geography by exploring the links between the two and their relations to scientific thought as a whole. Recent developments in geography have highlighted the need for a source of information on the philosophical basis of geography. Texts on the philosophical content of human geography are relatively common (Gregory, 1978; Kobayashi and Mackenzie, 1989; Cloke et al., 1991). Texts that serve the same purpose for physical geography are thinner on the ground.

Apart from the classic Haines-Young and Petch (1986) text on physical geography, there is little accessible to undergraduates that discuss how physical geographers think about and use philosophy within their work. Some physical geographers may regard this as a good thing, automatically stating the Chorley (1978) quote about reaching for their soil auger when they hear the term 'philosophy' (although the rest of the text beyond this infamous quotation is in favour of a more thoughtful approach to the subject). Physical geography has been and still is about 'doing' rather than theorizing in the passive sense of 'armchair' geography, an attitude with which many physical geographers interpret the term. This book is not intending to dispute the active and reflective nature of physical geography; in fact, this book is only possible because physical geography is a subject that views itself as being first and foremost about practice.

The activities of physical geographers, the practice of their subject, involves a vast array of philosophical decisions, even if most are implicit or seen as part of the tradition or training associated with a particular field. Importantly, physical geographers, as all good scientists, question what they are looking at; they do not necessarily take the world for granted as it is. They define, they classify, they probe, they question and they analyse. Why is this not seen as philosophical? The lack of explicit recognition of the importance of philosophy in physical geography may arise from its practitioners' reflection on what their human colleagues regard as philosophical. Human geography seems to have different underlying philosophies – seemingly many more than physical geography.

Human geographers seem to be forever embroiled in the latest debate about the nature of their reality, the construction of worlds, structuralist versus post-structuralist thinking, as well as the post-modernists' debates on the nature of geography. By extension, the number of books addressing explicitly philosophical issues in subdisciplines of human geography are legion by comparison to physical geography. Each faction seems to have its own clique and its own literature. Physical geographers are caricatured as having only one philosophy: the scientific approach. By implication, such an outdated and simplistic philosophical outlook cannot compete with the philosophical sophistication of human geography.

Physical geographers have not necessarily done much to correct this misconception of the scientific approach. Many physical geographers have preferred to retreat into their subject matter, dealing with the detail of their dating methods or the representativeness of their sampling methods, leaving the philosophizing to those who do nothing practical.

There has, however, been a set of wide-ranging philosophical debates that have occurred and are occurring over the nature of scientific investigation. Most physical geographers have been reluctant to enter into this debate in earnest. The 'physics envy' identified by Massey (1999) implies that only the hard sciences can develop philosophies which the 'softer' environmental sciences of geology and physical geography will then dutifully pick up and operate with. Fortunately, recent debates in physical geography with participants such as Frodeman (1995, 1996), Baker (1999), Demeritt (1996, 2001a), Rhoads and Thorn (1996a) and Richards (1990) have shown that physical geographers and their 'practical' approach to the environment have as much to contribute to the developing philosophy of science as do quantum physicists. Given the highly detached and impractical nature of theoretical quantum physics, it could even be argued that physical geographers can contribute more to the debate operating, as they do, within a discipline that has the physical environment at a more human scale as its subject matter.

Physical geographers have not been aphilosophical, but rather philosophy-shy. At the undergraduate level, the glaring exception to this has been Haines-Young and Petch's (1986) Physical Geography: Its Nature and Methods. Published in the mid-1980s and with an agenda that favoured critical rationalism as the scientific approach, this book provides a very general and readable introduction to some of the key philosophical questions that physical geographers need to address. At a more advanced level is von Engelhardt and Zimmerman's (1988) Theory of Earth Science. This advanced text gave a very Western European view of the philosophical basis of the earth sciences, and covers much of the same ground as the current text. The detailed level of the argument, however, was more complicated than might be expected of undergraduates unfamiliar with the idea of a philosophy in physical geography. More recent publications have taken a historical view of the development of the subject (e.g. Gregory, 1985) or have been collections of papers such as Rhoads and Thorn's The Scientific Nature of Geomorphology (1996c). The latter publication provides an interesting and thoughtful introduction to thinking in geomorphology appropriate for the postgraduate level. It does not, however, provide a coherent and integrated text that explores the recent debates about philosophy in physical geography.

In the nearly two decades since Haines-Young and Petch published their text, the nature of philosophy in science and, in particular, in environmental science has been debated and changed. Critical rationalism still stands as a beacon for understanding reality, but the nature of that reality has been subject to debate, as has been how environmental scientists should, or do, explore and explain it. The caricature of science as a monolithic philosophy with a rigid and singular approach to its subject matter is one that no longer stands up to scrutiny. In its place is a series of philosophies that share a belief in an external reality. Where these philosophies differ is in their claims about the ability of researchers to obtain an absolute knowledge of reality. Critical rationalism, critical realism and pragmatism have all been flagged in the scientific and geographic literature as being of relevance. In addition, the development of a more sociological view of scientific practice, as illustrated, for example, by the studies of the sociology of science from the Edinburgh school, has not really been addressed within physical geography. Likewise, the potential import of ideas such as Actor Network Theory (ANT) for understanding the context of practice within physical geography is relatively underdeveloped.

One of the things that has become clear is that it is the practice of a discipline that aids the development of its philosophy. Chorley may have been right to reach for his soil auger, but it is only because by the practice of a discipline that anyone can hope to understand its philosophy and, in turn, to understand how and why it changes.

IDEAS, CHANGE AND STABILITY IN PHYSICAL GEOGRAPHY

Ideas are central to understanding how physical geographers have observed and analysed the physical world. Ideas are not static, they change over time as well as from place to place. Ideas influence what we believe and how we understand reality. Ideas change internally, through the practice of a discipline, as well as externally through the social contexts within which physical geographers work. Differing ideologies, different philosophical approaches can influence the questions asked as well as the answers sought. Exploring this complex tangle of influences has often involved detailing the history of a subject. By following the 'great men' (and it usually was men in nineteenth-century academia) and their works, the key concepts from different periods can be traced and some sort of model of change developed. Often as important is the use of such a history to justify the current set of concepts within a subject.

This chapter briefly explores the history of physical geography to try to determine if there is a coherent current, accepted set of ideas that mould geographic thought. There may be no inevitability about the suite of ideas that form the focus of contemporary work. These ideas do not exist in isolation from particular philosophical approaches that have developed or been imported into physical geography. Philosophies can hide in the background, directing and constraining the type of questions asked, and the way in which they may be answered. It is important to provide at least some illumination of these philosophical approaches and their relative importance at different times in and for different parts of physical geography.

What are Ideas and how do they Change?

Ideas and theories are frameworks of thought with which people try to understand physical reality. A more detailed definition of theory will be covered in Chapter 2, but for this chapter an appropriate definition of a theory would be the core ideas or concepts that go unquestioned when a researcher studies the physical environment. From this core flow all the ideas and hypotheses that a researcher has about reality. The core ideas, and the central relations between these and derived ideas, form a tight, logical network. The dominance of vertical movement of land surfaces could be seen as an unquestioned assumption about how reality was in geology at the turn of the twentieth century. From this central, core idea other ideas flowed that explained the distribution of species that did not contradict the evidence on distributions available. Rapidly rising and submerging land bridges were mentally constructed to explain the range of distributions of fossil creatures between South America and Africa. As this central idea was replaced by the acceptance of horizontal movement, continental drift and plate tectonics were developed as an alternative explanation for these distributions.

A number of authors have tried to generalize how ideas change. Three 'big' ideas are progressive change, paradigms (Kuhn, 1962, 1977) and research programmes (Lakatos, 1970). Progressive change states that ideas in science change incrementally and slowly. Current researchers build upon the work of others, correcting their errors until finally, at some vastly future unknown date, everything is understood and reality, as it really is, is known. Accumulation of establishable facts counts as progress of knowledge. Once a fact has been discovered it remains the same for all time. Current research is just adding to and extending existing knowledge. Unfortunately, understanding of reality does not seem to follow a nice, steady and progressive path. There are differences in the rate at which ideas change. Some may suddenly appear apparently from nowhere and some seem to disappear, never to be mentioned in polite scientific circles again, such as racial superiority and acclimatization. This view assumes that earlier researchers understood reality and derived the fundamental laws and relationships that later workers have merely embellished.

Kuhnian paradigms were developed to explain how ideas in science appeared to undergo rapid and revolutionary changes. Kuhn's prime example was the switch from the Greek to the Copernican model of planetary motion, with the sun replacing the earth as the centre of the universe. Although much criticized, particularly by critical rationalists, the idea of a paradigm has taken hold within geography and been used to describe various perceived movements, fads or schools of thought. Paradigms, in other words, have been used to describe just about any group that is perceived to have a common and coherent set of ideas about the world, whether that group realize they have these or not. Kuhn never really clarified or defined what a paradigm was and so there has been wide interpretation of what he meant by the term.

Figure: Science as a progressive pursuit of absolute truth.

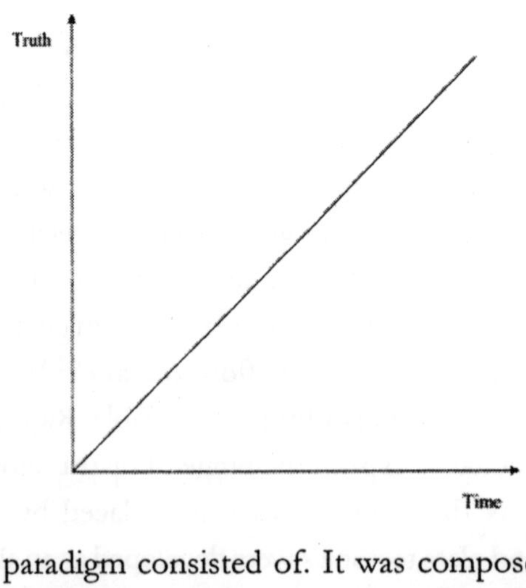

He did outline what a paradigm consisted of. It was composed of a disciplinary matrix of tangibles and intangibles. Tangibles were examples of good or appropriate practice within a discipline: classic textbooks and classic experiments. These provided concrete illustrations of how to go about research and what was acceptable practice. In addition, tangibles included objects such as test tubes, laboratory benches and the like, which were viewed as essential to the practice of the discipline. Intangibles were the unwritten rules or conduct expected in a discipline: the social norms of behaviour and accepted evidence as dictated by the peer group of the discipline.

Change in this system was brought about not by the progressive and careful alteration of existing ideas in the light of new evidence or new techniques, but by the wholesale overthrow of old ideas by a complete and coherent set of new ideas. Kuhn suggested that most scientists undertake what he described as 'normal' science. They work diligently at their benches using standard techniques to prepare standard solutions, to undertake standard tests. They ask only questions that they know they can answer using these techniques. There is little to challenge the established view of how reality is. This set of concepts and practices would make up the existing paradigm. Slowly Kuhn believed that facts or observations would accumulate that the existing concepts could not explain. Eventually, the errors would become so pronounced and interfere so greatly with the operation of 'normal' science, and so restrict the questions that could be answered, that there would be a complete switch from the existing ideas to a new set of ideas that explained the 'problem' facts.

Figure: Science as a series of paradigms changing in a revolutionary manner towards a relative, but unknowable, 'truth' – or at least that is what we would like to believe. In fact the lines could go in any direction relative to absolute truth, the way reality really is.

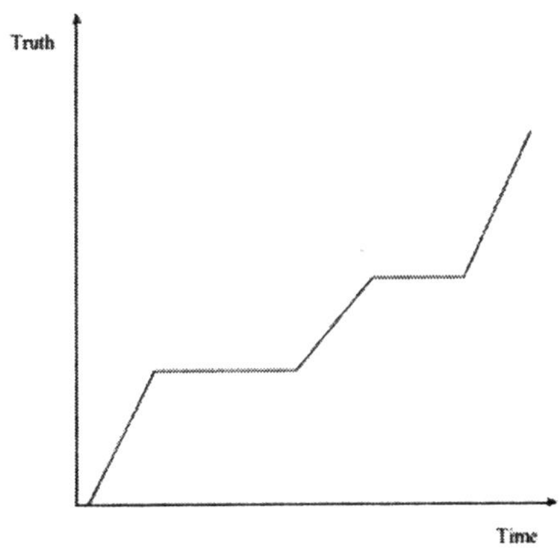

The paradigm switch was rapid and complete. 'Seeing' the world through one set of concepts, the paradigm, meant that you could not see it through the other set. There can only be one paradigm at any one time, and seeing the world through that single paradigm excludes the possibility of all other views of the world. There is nothing to guarantee, however, that your particular view of the image is the correct one. This is because switching from one paradigm to another means a complete change in what is viewed as evidence and even what is believed to exist in reality. This means that there can be no argument over the facts because the facts are different in each paradigm; indeed, what is viewed as a valid argument may even differ between paradigms. Paradigms need not operate across time alone. There may also be spatial differences in what is accepted as a valid view of reality. How can you decide between paradigms? Kuhn argued that although it may be clear with hindsight that paradigm A was a better explanation of reality than paradigm B, at the time the choice could not be made on any rational basis.

Instead Kuhn turned to his set of intangibles and suggested that the choice between paradigms was based on sociology rather than 'hard', factual evidence. Peer pressure decided which paradigm to accept – not facts. This was the part of the theory that critical rationalists found particularly hard to stomach. It implied that there was not necessarily any movement of ideas towards a true understanding of reality. If paradigm choice was socially based, then the new paradigm may not get you closer to how reality really is. Despite the new paradigm eliminating some of the errors or unexplained facts under the old paradigm, the improvement has been at the cost of all the concepts and understanding embodied in it. There can be no rational discussion about the relative merits of either paradigm as there is no common basis for such a discussion. Kuhn's vision of science involved young, dynamic researchers overthrowing the treasured and established views of an old guard. These researchers worked on critical experiments designed to disrupt the basis of the old order.

Research into critical questions posed to undermine a paradigm was contrasted with the mundane 'normal' science of the established order. But does this romantic view of scientific progress match what actual happens? Lakatos thought that Kuhn's ideas did not match his experience of science and how it changes. Lakatos was a strong supporter of a rational basis for choosing between competing sets of ideas and searched for a way to incorporate a critical rationalist approach into this process of choice. Lakatos divided sets of ideas into research programmes. Each programme had a central set of core ideas, the heart and soul of a set of theories. These directed much of the work the researcher undertook, constraining what questions to ask, how to ask them, and the techniques and explanations that were valid in any answer (the positive heuristic).

In addition, the research programme also identified the questions that researchers should not ask, techniques and modes of explanation that were unacceptable (the negative heuristic). The core theories were never directly tested; they were kept away from direct assessment. Instead a band of auxiliary hypotheses formed the basis of the active use of each research programme. These were testable statements derived from the core theories. These were the elements of the research programme that researchers used in their scientific lives. Disproving one hypothesis did not bring the whole of the research programme into question.

The nodes represent axioms in the core and hypotheses in the outer circle. These nodes are linked by vertices that represent relationships between nodes. The progressive research programme has a rich and internally highly connected core of axioms. These generate a range of hypotheses that can even interact with each other to generate 'second'-generation hypotheses (or more). The degenerative research programme has a limited set of core axioms that generate few isolated hypotheses. That hypothesis was rejected and some other hypothesis constructed using the existing central theories.

Lakatos suggested that choosing between different research programmes became a matter of rational choice. Research programmes that continually generated new hypotheses, novel ways of questioning reality and which were generally not disproved, were progressive programmes. Research programmes that were stagnant and did not advance new ideas or explanations for reality were seen as degenerative. When there were two competing research programmes researchers switched to the more progressive research programme. Some individuals might take longer than others to see that one programme was more progressive, but eventually all would make the same rational choice. In Lakatos's view of science there was elimination of errors and change of world-view, but each change brought the researcher, or rather the scientific community, closer to the truth, to the way reality really is. Unfortunately, it would never be possible to prove that reality had been captured absolutely by the theory, only that the theories of a particular programme had not been disproved.

Both paradigms and programmes have been used to describe changing concepts in physical geography. Neither is perfect and each has a number of problems. Adhering to paradigms would imply that at any one time a single, all-embracing set of ideas would direct the researchers in a subject. This is rarely the case. Similarly, adherence to research programmes implies that central sets of concepts are somehow immune from testing and unaffected by the failure of hypotheses derived from them. Such an ivory tower existence for privileged theories is rarely sustained. Some authors have suggested that there are general currents in thought and that theories follow these. Researchers are moved by fads and fashions as much as anyone else and the flavour of the month has as much relevance in describing academic work as any other. Identifying how such fads are translated to theories or a mechanism by which changing ideas occur within fads has not really been tackled. A significant problem with both frameworks is that they do not really address how and why certain ideas seem to survive as general, favoured theories change, whilst other ideas are rejected. The stability of some concepts, the recognized, but malleable nature of others and the outright rejection of still others is not addressed.

How ideas change is still open to debate despite the models presented here. One of the important points to bear in mind is how each model views the ultimate goal of research. If you believe that knowledge is progressive and developing towards a final complete understanding of reality as it is, then Kuhn's paradigms, with their sociological basis, are unlikely to be attractive. If you believe that all knowledge is relative then paradigms may be more attractive. Lakatos seems to sit on the fence by accepting that ultimate absolute knowledge is unattainable, but then implies that there is a progressive march to this goal that helps to decide between research programmes. This would suggest an underlying belief in the movement of ideas towards representing and explaining reality as it really is, even if you can never be certain you have arrived.

Is there a History of Ideas in Physical Geography?

Physical geography is a diverse subject with many, shifting subdisciplines vying for status. Outlining a single history of such diversity is both problematic and misleading. Detailed histories of ideas and their development can fall into the trap of reading the past by the light of the present.

Trends, unbeknown to workers at the time, are identified with contemporary certainty and clarity. Such histories can lapse into justifying the present rather than trying to understand the past concepts in their appropriate contexts. Often these histories are involved in justifying current practice or concepts by referring to a long and hitherto neglected (or unknown) history. Such reinterpretations of the past often provide intellectual justification for the present by an appeal to the authority figures of a noble past.

The first thing to note in looking at the history of physical geography is that there is not a single history of physical geography. There are many possible histories that can be constructed from the information available.

Judgement of a single narrative as the 'correct' one implies a very restricted view of what is important in physical geography and how it came to be. If there are many possible histories, how can anyone identify what is important from the past? It is essential to have a clear idea of why you want to look at the history of physical geography.

This objective will determine what you consider to be important. In this book, my purpose in reviewing the history of physical geography is not to provide a definite history. Instead there are three key concerns that I want to highlight as being present in the study of the physical environment: universality of explanation, a concern with the empirical, and a concern with explaining change or stability. Tracing how these concerns have been expressed and the strategies for answering them is the focus of my thumbnail history. Even this limited ambition is fraught with difficulties.

Delving into the history of physical geography requires identifying the subject in the past. This can be a major problem.

To identify individuals, and what they did as physical geography, requires that physical geography itself be seen as a legitimate subject area. As an academic subject, physical geography requires professional academics, academic institutions within which they work, journals within which papers in the field can be published, and so on. The subject requires extensive legitimizing networks of relations between academics, between institutions and other organizations such as publishers. These networks were not necessarily around in the past nor in a form that would permit the identification of physical geography as it is understood today.

A more fruitful approach might be in identifying the central subject matter, and defining the histories of physical geography that way. There is a problem with this approach as well. It assumes that the same subject matter has been studied throughout academic disciplines over time and space. Although it may be possible to identify the ancient Greeks as studying climate, were the phenomena they studied the same as contemporary or even Victorian academics would consider climate? Were they collecting information on the same phenomena?

Even if there was some commonality in the nature of the phenomena under investigation, is it possible to abstract it from the whole intellectual framework of which it is a part? Newtonian mechanics is usually viewed as a key element of the foundation of modern science, and Newton's 'modern' approach of experiment confirmation is viewed as exemplary. The alchemy aspect of Newton's work, and the manner in which this was intertwined with his 'acceptable' science, is rarely mentioned (White, 1998). For Newton, both aspects of his work were vital; he made no distinction between what we would deem 'proper' science and 'pseudo-science'.

Another possible approach is to look at the 'great' figures in the subject and define your histories by reference to what they thought and did. Although it may not be possible to identify them as physical geographers (remember the term might not have existed), at least their work survives to be reinterpreted as the appropriate subject matter of contemporary physical geography.

Appropriating figures such as Darwin, Wegener and Huxley as influential in developing ideas in physical geography neglects the multidisciplinary or even interdisciplinary nature of their work. The environment these individuals worked within was different from the present and the subject areas they recognized would not have necessarily included what contemporaries would view as the 'core' of physical geography. It is also likely that other disciplines will claim these figures as originators in their subjects as well.

Appropriation of 'great' men will involve teasing their relevant work from the rest of their intellectual endeavours, a division they would not have made. It could be argued that even this approach creates histories that mirror contemporary concerns, rather than really reflect the work of these individuals.

Likewise, focusing on great men (and it usually is men in the gender-closed Victorian intellectual world of geography) neglects the contribution of 'lesser' figures who undertook practical work that might be described as geographical. The army of surveyors, for example, who undertook the surveying of the British Empire, trained and developed methods for describing and collating spatially referenced information (Driver, 1992; Collier and Inkpen, 2003).

The data and analysis undertaken by these forgotten figures were vital in developing imperial policies and developing research methods, yet it is often regarded as a lesser achievement than abstract theorizing. Contemporary intellectual snobbery can cloud what is deemed relevant and what is deemed unworthy for a subject's history.

Most Anglophone histories of physical geography tend to focus on the Anglo-American experience and definition of the subject. This limits the types of histories that can be written. Although it could be argued that this impression reflects an inability to read well in other languages, it also reflects a belief in these areas that they represent 'real' geographic thought.

Some histories give nods to early work by Portuguese navigators or Arab intellectuals, but from the eighteenth century onwards the Enlightenment and scientific revolutions bring with them a view of 'proper' science finally being done.

For 'proper', read 'science' as we Western Europeans would understand it today. Such views are also translated into contemporary global surveys of geography. AngloAmerican-derived topics and ideas occupy centre stage; other cultures are relegated to minor interests. It could be argued that this concern with Anglo-American physical geography reflects a wider pattern of change since the eighteenth century.

Development of a capitalist world economic system, some would argue, has marginalized and stifled any independent intellectual development of indigenous cultures. It may be, therefore, not surprising that the only 'acceptable' histories are those with a Western, and specifically an Anglo-American, flavour. Although acceptable to whom – and why them – is another loaded question that is too complicated for consideration here. The possible bias in the construction of histories, and the reasons for it, are rarely touched upon in histories of physical geography.

All the above might imply that there is no point to trying to even produce a historical view of physical geography. There is no single history, there is always the problem of interpretation and there is always likely to be subjective bias.

There is, however, still a case to be made for sketching even a flawed history of physical geography. Identifying ideas within their context and tracing how some ideas have evolved may help in understanding why some ideas remain relatively stable and others fade into obscurity. Stable ideas are those that appear to retain their basic tenets and produce results that are acceptable to the academic community of the time.

The Davisian cycle of erosion, for example, could be viewed as an idea that no longer provides acceptable results upon which contemporary geomorphology can operate. It does not provide a framework for asking or answering questions felt to be appropriate to practising geomorphologists. The cycle of erosion becomes a distinguished or obscure, depending on your viewpoint, element of the history of the discipline.

Plate tectonics, on the other hand, has evolved to be a major and overarching framework for posing and answering geomorphic questions about long-term landscape change. From the 1960s, the general idea of plate tectonics has been developed and refined to create a complicated range of subdisciplines, each with their own particular criteria for evidence and working practices. Tracing the decline of one idea and the rise of another brings to the fore both the shifting nature of what is viewed as real and the surprising stability of much of the content of physical reality.

What are the Important Concepts in Physical Geography?

The three themes developed as central to the history of physical geography are not independent. The quest for universality in explanation has been driven by the need to explain landscapes and their changes. Likewise, identification of the changing or stable nature of the physical environment has been aided by increasingly complex empirical information.

The search for universality can be viewed in terms of both an overarching theory as well as an integrated method for analysing the physical environment. Huxley's Physiography (1877) is viewed by Stoddart (1975) as an important exercise in forming an integrated approach to the subject matter of the physical environment.

Stoddart sees Huxley's approach as being in a direct lineage from the conception of the physical environment of Kant and Humboldt nearly a century before. Physiography starts at the micro-scale, with the familiar, and works outwards to the macro-scale and unfamiliar.

In so doing, Huxley develops an explanatory framework that begins with identifying and classifying causes at the local level and then working from these familiar, or, as he would have it, 'commonsense' illustrations to the wider picture. Cause-and-effect relationships are built from individual experience, emphasizing the empirical over the theoretical, but applied to increasingly larger spatial and temporal scales. The local is then viewed as part of this wider context and understanding of the local requires this wider context.

Huxley's approach had much in common with the 'new' geography being developed by Mackinder (1887) in highlighting the importance and significance of integration and synthesis of information, both physical and human. Much as with Mackinder's new geography, the high ambitions of physiography were never really fulfilled in developing physical geography. Davisian geomorphology overtook Huxley's physiography as the main explanatory framework of geomorphology, and the dream of an integrated study of the lithosphere, atmosphere and biosphere was increasingly replaced by specializations.

From the perspective of the early twenty-first century the failure of both Huxley and Mackinder to have the influence that their holistic and integrative approaches should have had, or were supposed to have, is of interest. These two figures were influential in their own right and yet their ideas, that strike several chords of recognition, failed to develop as the core of the subject. This illustrates the first problem in outlining a history of physical geography. Both Huxley and Mackinder are viewed as using a Darwinian framework for their ideas, as being good disciples of Darwinian thought. Does their failure, then, reflect a failure of Darwinian thought to influence physical geography? Looking at Stoddart (1966) the answer would clearly be no. According to Stoddart, Darwin's views of ecosystems, the integrative nature of the environment and the significance of change have permeated geographical thought.

The central core of Darwinian thinking (Gould, 2000) – that of non-directional evolution through random variations in organisms – has had a lesser impact upon geographical thinking. It is only specific interpretations of Darwinian evolution that have affected geography, not the detailed theory as outlined by Darwin in 1859. The first problem, then, is not to fall into the trap of assuming that there was a single or universally accepted interpretation of a potentially integrating concept such as evolution around in the past.

A vital rallying point in the search for universality in explanation, as noted by Goodman (1967), was the concept of uniformitarianism. The term, usually attributed to Charles Lyell (1833), has been paraphrased as the idea that 'the present is the key to the past', the general idea being that by observing the present in terms of forms and processes it should be possible to apply this knowledge to the past. Everything that happens now should have happened in the past. Gould (1965), amongst others, has pointed out the major problems with maintaining a substantive rather than methodological view of uniformitarianism.

Gould (1965) defined substantive uniformitarianism as the concept that the rates of operation of geological processes have been constant or uniform throughout time and space. Upholding this view would imply, for example, that there has been a constancy in the number and type of events causing the landscape to change, and so in the magnitude and frequency of the processes causing change. This view of uniformity led Lyell to an extreme cyclical view of landscape and biological change (Kennedy, 1992; Gould, 1987) in which he could seriously envisage a cycle to life with extinct species eventually being recreated.

Methodological uniformitarianism referred to the assumptions that underlie any study of the past and present in a historical and empirically based science.

First, the laws that operate to produce change or stasis are assumed to be constant in both space and time. This is a vital assumption as it means that explanations for change and stasis in one location can be applied to other locations and other time periods with confidence that they will operate in a similar manner.

Second, it is assumed that explanation of the present (and indeed the past) does not require any invocation of unknown or unknowable processes. This is the assumption that uniformitarianism safeguards most saliently. At the time of its development, uniformitarianism was a concept set up in opposition to the prevailing attempt to explain all of nature, and more importantly what was not understood about nature, by final reference to an overarching deity. This final source of all explanation could be invoked to explain why the world was as it was; there was no need to seek a material, earth-based explanation – reference to the guiding principles of a deity was sufficient.

This should not, however, be interpreted as meaning that the explanations offered were simplistic or unreasoned. On the contrary, the most learned brains of the time subscribed to this view and debated the 'correct' interpretation of evidence in a logical and reasoned manner. It is just that the basis of their logical system permitted explanation by reference to a deity as much as explanation by reference to observable processes and agents. Hutton and Lyell used purely material, empirical observations and reasoning restricted to what they could observe and infer about material reality to construct explanations of features in the landscape.

This was a radical step and at a stroke it removed catastrophism from scientific explanation. This was important as it meant that dramatic biblical events, such as the flood, could no longer be used to explain observations such as extinctions or the unconformity of sediments. Likewise, it meant that observations no longer had to be shoe-horned into a temporal sequence that was restricted by biblical events.

The empirical 'facts' now had greater weight in a scientific argument than the enforced explanatory framework. Time, as the arena within which processes operated, was no longer limited to the scale of biblical events. Out of this release, however, another constraint was developed: the idea that there could be no sudden changes in process rates or the occurrence of processes not currently active. This meant that explanations that made use of what would today be considered as extreme events, such as meteorite impacts, were excluded from the explanatory framework.

The important ideas of uniformitarianism lie in the concepts from which it is derived. Uniformitarianism assumes that everything observed in nature can be explained, the principle of causality as an overriding universal principle enters into explanation. Everything observed is capable of explanation, everything observed has a cause.

The cause is not based in some unobservable and hypothetical other world, in some omnipotent deity. Causes are to be found in what we can observe in the here and now. Cause and effect only have recourse to the material world, not to the supernatural. Uniformitarianism is in this sense, as Goodman (1967) suggests, the simplest explanation for the physical world and how it changes. From this one idea, applied consistently and continuously in physical geography, the importance of the empirical rather than the theoretical, of the observable rather than the unobservable, and of the location of causality in the material present are to be found.

Of vital importance in nineteenth- and early twentieth-century thinking about the physical environment was the use, or rather interpretation, of the concept of Darwinian evolution. Darwinian evolution could be seen as representing the application of uniformitarian principles to the organic realm. In this manner, it is another attempt to provide an overarching theory to explain the natural world. Evolution is, however, more than just Lyell amongst flora and fauna.

Bowler (1983a, 1983b) has written on the development of the concept of evolution before and after Darwin, and Livingstone (1984) has illustrated how the concept of evolution was interpreted in a particular manner by American physical scientists.

Darwin had expounded a theory of change in the organic realm by means of random variations in individuals. Individuals competed for resources and, where the random variation was beneficial in this competition, successful individuals produced more offspring. The offspring had the beneficial variation that became a dominant characteristic in the species over time. From initially small differences between individuals multiplied over the vastness of geologic time, species became differentiated. The random variation upon which this process of natural selection was based was non-directional. This means that the course evolution takes is not predictable; uncertainty and randomness lie at the heart of Darwinian evolution.

Central to North American interpretations of evolutionary thinking within physical geography was the interpretation of evolution used by W.M. Davis in his cycle of erosion. Essentially Davis used the concept of progressive change through predefined stages as the basis for his view of evolution.

This view was not the one expounded by Darwin involving random variation and non-directionality. Davis's view, however, was firmly within a North American tradition of orthogenesis and progressive change as developed by Hyatt and, within Europe, by Haeckel. Haeckel viewed changes in evolution as occurring by the addition of stages to an existing established sequence. An embryo, in his view, went through all previous evolutionary stages, with a new stage, that of humans, added at the end of the embryo's development.

The same sequence of stages could be observed in embryos of different species, but for the 'lower' species this happened without the addition of more 'advanced' stages. So whilst a human embryo might be expected to go through all stages from fish through amphibians to reptiles to mammals, a fish would not have these additional stages visible in its embryonic development.

Gould (1977) uses the term 'terminal addition' to describe Haeckel's vision of evolution. Gould also notes that this viewpoint was compatible with the Lamarckian view of the inheritance of acquired characteristics and that Haeckel did indeed subscribe to the transfer of such characteristics between generations. Haeckel's views from biology have a resonance with the ideas of Cope (1887) and Hyatt (1897), the latter a palaeontologist working initially under Agassiz in the US. They believed that although there may be a preordained sequence of stages, organisms could acquire characteristics as well and pass these on through generations.

Central to Cope and Hyatt's work was the idea that the stages that organisms went through in their development could be accelerated or slowed down. Progressive evolution saw stages being condensed further and further into the early stages of development, allowing more time for the addition of acquired characteristics at the terminus of development. Retrogressive evolution saw stages being slowed almost to a stop so that stages are lost before development ends. The former they termed the law of acceleration, the latter the law of retardation.

Their significance lies in the possible influence they had upon the development of evolutionary ideas in the US. Rather than the non-directional, random and chancy process of selection for natural variations in characteristics of an organism independently in each organism's lifetime, Cope and Hyatt strove for an almost opposite view of change. Within their scheme, evolution was about movement through a series of predefined stages with acceleration or retardation adding or losing stages through the sequence. This is the form of evolution that should be borne in mind when discussing Davisian evolution and the development of ideas from it.

Any landform and any landscape had to pass through the stages set by Davis in his cycle of erosion (1899). Any landform could be classified by its morphology to a particular stage in this predetermined evolution, through youth, maturity and old age. Explanation was contained within the theory, within the evolutionary model used.

Observation merely confirmed the position of a specific landscape or landform within this scheme. Stage became the explanation rather than what happened in the landscape. Significantly, however, the stage model highlighted the importance of identifying and providing an explanation for stability and change in the landscape. Both types of behaviour could be explained by reference to the same explanatory framework. Change was an essential feature of evolution and stability was illusory, based on the inability of humans to perceive alteration at a time-scale of relevance to the landscape.

These characteristics of the cycle of erosion provided it with a scientific veneer, particularly as it explicitly used evolution, the scientific theory par excellence in the nineteenth century. Bishop (1980) and Haines-Young and Petch (1986) have pointed out the deficiencies of the Davisian model as a scientific theory, but its perception as a source of progress in explanation in physical geography in the nineteenth century, and as a guiding framework for geomorphic thought in that period, is generally not questioned.

Davis's interest was, however, focused at the macro-level only, although his quest for explanatory frameworks admitted to the operation of the same processes, the same models, across all scales as in his work on tornadoes and their similarity to the spiralling of galaxies. In this manner, Davis believed he had found an overarching and universal mode of explanation for phenomena in the physical environment. It would be inappropriate, however, to view nineteenth-century physical geography as just about Davis and geomorphology.

Even at the scale of the landscape, however, the Davisian model was not the only available theory: Penck had also developed a model of landscape development. Penck's model has been reinterpreted for the systems era as an early attempt to apply a coherent systems framework to the diverse factors affecting landscape development.

Although Penck would not have understood his theory in those terms, his use of factors and their relations to define landscape development does seem capable of easy translation (e.g. Thorn, 1988). A significant difference between Penck and Davis was the nature of changing slope form in each theory. In Davis, slopes declined, whilst in Penck slopes were replaced. In slope decline, denudation processes remove material from the top of a slope at a faster rate than they remove material from the base of the slope.

Gradually, the slope as a whole declines in angle. In Penck's theory, the base of the slope becomes covered with debris denuded from the top of the slope. Gradually, the base of the slope is replaced by a debris slope of constant angle.

In the 1950s King proposed a third type of change: slope replacement. In this theory, slope elements remained constant (waxing, waning, free-face and pediment), only the length of the pediment slope increased over time as debris from the other slope elements reached it and extended it. All other slope elements remained at the same angle throughout their existence.

Although differences in the three theories can be partly related to the different locations of their production, as noted by Thorn (1988), the three theories do illustrate a commonality in their use of time. All three are time-dependent; that is, all three presume that change will be progressive over time and the nature of the form of the landscape will change over time. Denudation chronology has been viewed as a mode of explanation directly linked with the Davisian cycle of erosion and often touted as the dominant mode of explanation in physical geography until the late 1950s.

Denudation chronology attempted to reconstruct the past evolution of the landscape from the sequence of denudation surfaces still visible in the landscape. Evidence for this relative dating of land surfaces was to be found in the morphology of the landscape, its sediments, from absolute dating techniques of portions of the landscape and from an understanding of the processes of landscape development (Gregory, 1985).

Any evidence collected was assessed within a framework of stages of development of the landscape. Evidence did not stand alone, but had to be interpreted within this framework. A great deal of the methodology and theoretical debate over the identification and presence of entities such as planation surfaces centred, as Gregory noted (1985), upon a few 'classic' studies such as Wooldridge and Linton's (1933) interpretation of the South Downs. Even the advent of the statistical analysis of data did not initially diminish the appeal of this approach to landscape study, as trend-surface analysis was employed to aid the identification of 'real' surfaces in landscapes.

Another alternative set of theories concerning landscapes and landforms presumed that change could be understood as time-independent. In these theories, forms did not change over time, there was instead a steady balance between variables that produced forms that could not be placed in a temporal sequence.

Gilbert's concept of dynamic equilibrium (1877) and Hack's model of the same (1960) are classic examples of this approach. Both Gilbert and Hack viewed spatial variations as more important than temporal variations. Spatial variability replaced temporal variability as the dominant feature of these theories. In this view, the landscape is in balance with current inputs and so there is no temporal memory in the landscape.

The only possible sources of variation in the landscape are not the different stages in development or even past histories, but instead spatial variations in the nature of and inputs into the landscape. Although the exclusion of the 'past' as a source of variation in the landscape is a problem in time-independent concepts, they do form a basis for trying to understand the landscape as a product of contemporary processes. These models bring landscape-forming processes to the time- and space-scale of the investigators and detach the landscapes from their unknowable (in human terms) pasts. They provide, in other words, a platform for examining the landscape at spatial and temporal scales relevant for applications such as engineering.

The above two approaches to landscape study are very Anglo-American in their orientation. There were other theoretical frameworks within which landscapes could be explained. Two of the most significant on continental Europe were climatic and structural geomorphology. Climatic geomorphology, as the name suggests, viewed climate as the overarching explanatory framework for landscapes. This is the view that climate controlled the characteristics and distribution of landforms. Climate is the controlling factor in landscape development. Climate resulted in a distinct latitudinal zonation of landforms, with an additional similar variation for the effect of altitude. There was a consistent causal relationship between the two, between climate and landscape, based on the persistent and consistent presence of a set of landforming processes unique to each climatic environment.

A zonation of landforms on the basis of climate was also developed by Tricart and Cailleux (1972) in France. The climatic zones, and indeed continents, were not static and this was reflected in the landforms found in different regions. Landforms and landscape could reflect the action of past climates, and so the simple latitudinal zonation had superimposed 'historical' patterns of climatic influence.

Linton's (1955) model of tor development on Dartmoor was a classic example of the application of climate and its change as the explanation for the development of a specific landform. Linton hypothesized a two-stage climatically controlled series of processes.

First, the core stones of the tors were developed through the deep weathering of granite under tropical conditions. Large core stones were produced at depth where weathering was less intense and smaller core stones near the surface where alteration was greatest.

The regolith remained around the core stones until removed by an episode of periglacial activity. This climate was more effective at removing weathered material, and with this removed the core stones formed disjointed masses on hilltops and valley sides. The idea of 'natural' zonation of the physical environment, particularly as controlled by climate, was applied not only to geomorphology. Herbertson (1905) developed his idea of natural regions based on the distribution of biota under the influence of climate on an ideal continent.

Peltier's (1950) version of the importance of climate related more to the relationship between process and climate than directly to landforms. Peltier insisted that climate controlled the ability of processes to operate.

A particular combination of temperature and precipitation would invariably result in the operation of mechanical or chemical processes of denudation or a particular combination of both. Although this vision of geomorphology was developed, and still influences geomorphology on continental Europe, it found little favour in the UK, as noted by Gregory (1985, quoting Stoddart, 1968; Derbyshire, 1973; Douglas, 1980).

The main criticism seems to have been that the idea of climate affecting landform and landscape development was perfectly acceptable, but the idea that it was the major theory for landscape development was not. The view of a specific climate resulting in a specific landscape was seen as too simplistic and not able to cope with the complexities of interpretation that structural differences and past histories imposed on it.

These early trends in the classification and quantification of natural phenomena tend not to be mentioned in 'standard' histories of physical geography. The reasons why are unclear. One possible reason is that these measurements were often made without the use of an overarching explanatory framework within which to place the measurements.

Gregory (1985), for example, notes that Kuchler categorized the early twentieth century as a period of data collection and accumulation during which there was no co-ordinating or overarching theory with which to explain the patterns observed. Within subdisciplines such as meteorology, however, this was not the case as theories of atmospheric circulation and weather patterns were available to co-ordinate and explain temperature patterns, windflows and other climatic parameters. The recent book by Solomon (2001), for example, illustrates that there were accepted explanatory frameworks based on detailed theories for weather patterns even in the data-poor Antarctic region at the end of the nineteenth century.

These theories were used to predict the likely behaviour of weather patterns and for planning expeditions, although the extreme and unpredicted weather of 1912 resulted in the tragedy of Scott's failed expedition. The early twentieth century would see these patterns and classifications extended by the likes of Koppen, and later by Thornwaite (1948), but the numerical basis for these schemes was developed in the nineteenth century.

By the 1960s the questions that physical geography was asking of the environment had altered, and the explanatory frameworks used had to evolve as well. Davis's cycle of erosion focused research onto the megascale, onto questions of landscape development over eons.

Pressing questions of engineering concern, such as how is a landslide going to behave if a road is cut through it, were not answerable within this theoretical framework. Whether this type of question was appropriate for this framework, and by implication an appropriate question for geomorphology, was not, it seems, a concern. The nature of questions in physical geography had altered since Davis posed questions on a 'grand' scale. The quest for relevance and the need to justify their academic existence meant a focusing of geomorphology on questions of relevance to society.

Whether these questions were questions of relevance to everyone or to particular groups in particular locations has not been made clear in the histories of the subject. Of central concern, however, was the recognition that most questions concerning landscape on a human spatial and temporal scale were just unanswerable as well as unaskable using the Davisian framework.

Identifying a planation surface did not help in identifying the stability of a specific hillslope, nor did the identification of a section of the landscape as mature help in identifying and predicting rates of change. The nature of questions of interest to geomorphologists had altered, and so their operating framework had to change.

Other parts of physical geography had long discussed issues at scales other than the regional and geologic. Climate studies, for example, had integrated short-term measurements of changes in variables such as temperature into explanatory frameworks at a range of scales. Prediction of weather patterns relied on the rapid collection and processing of information within changing models of how the atmosphere operated. Small-scale and large-scale phenomena were both recognized and modelled using the same physical laws. Aggregation of behaviour was needed to predict at the regional and global scale, but the principles on which the models were based were similar.

Ecological studies had likewise developed a scale-independent framework for analysing and integrating information. The ecosystem concept had been forcefully put forward by Tansley (1935). The ability to identify, classify, define and relate entities within the landscape at any scale made this form of explanation an extremely powerful tool for understanding.

Geomorphology overcame the scale problem of the Davisian model by sidestepping it. As Haines-Young and Petch (1986) noted, the solution was not a new theory, it was an alternative means of interpreting the information that already existed. The new interpretation could not be tested directly, but its explanatory framework offered a means of ignoring the temporal and spatial scales of Davisian focus. Schumm and Lichty's classic paper (1965) overcame the Davisian framework by restricting its scope of operation to a specific type of time: cyclical time. This was the longest temporal scale they envisaged.

At this scale only certain variables, as identified by Davis, were of significance in explaining landscape development. In other words, there were only a limited number of variables that changed at this scale.

There were, however, two other temporal scales: steady-state and dynamic equilibrium. At these two scales the variables that were significant for long-term landscape development could be treated as if they were constant, and so were unchanging. Other variables, considered insignificant in cyclical time, became variable, and so significant at the other two temporal scales. Initial relief, for example, is significant as an independent variable in cyclical time but irrelevant as a variable at other time-scales. Drainage network morphology is a dependent variable at cyclical and graded time-scales but an independent variable at the steady-state time- scale.

In this manner Schumm and Lichty provided the beginnings of an explanatory framework that retained the Davisian view but which permitted other conceptual frameworks for explanation to be applied at other temporal, and, by implication, spatial scales. Systems analysis developed rapidly as the integrative explanatory framework of physical geography.

As important, however, were the 'new' questions that this approach made legitimate – questions already being asked and answered about processes of change and their rates. Although the importance of process for understanding geomorphology had been recognized (e.g. Strahler, 1952), and even attempted as a basis for classification in climatic geomorphology, the seemingly immature development of the subject relative to the 'hard' sciences seemed to condemn geomorphology to a 'mere' description of forms.

Identifying and understanding processes and their rates of operation were essential, even to the development of the Davisian explanatory framework. If process rates could be identified from current processes and agents then extrapolation and postdiction could be made and the length of Davisian stages calculated. Changing the scale of study, however, permitted collection of empirical data on processes and so the immediate assessment of the operation of environmental systems.

A spate of publications in the 1970s and early 1980s (Huggett, 1980; Trudgill, 1977; Chorley and Kennedy, 1971) all illustrate the acceptance of process studies as a key feature of geomorphology. This is not to imply, however, that such views were not around before the 1960s and 1970s. Gilbert's work on the Henry Mountains (1877) and dynamic equilibrium was an alternative approach to Davis's historical geomorphology.

The development of the physics of sand movement by Bagnold (1941), the association of mathematical modelling and hydrology by Horton (1945), the study of mass movement by Rapp (1960), as well as the research of Strahler (1952) and Schumm (1956), were all examples of process work in geomorphology. Gregory (1985) assesses the publication of Fluvial Processes in Geomorphology (Leopold et al., 1964) as the most influential publication in the 'new' field of process studies.

These authors were working within an academic context where process studies were not unknown and where systems analysis was beginning to take hold as an alternative explanatory framework to the cycle of erosion. Their work was important, but it illustrates a trend rather than marks a discontinuity. It could also be argued that Gregory was talking about an academic audience that was relatively small and insular.

The textbooks mentioned above were aimed at specialist undergraduate courses, presumably taught by academics who agreed with the approach taken. As important, if not more so, are texts that were written at the time for a wider audience. In this context Cooke's book Geomorphological Hazards in Los Angeles (1984), as well as Cooke and Doornkamp's Geomorphology in Environmental Management (1974), were early examples of attempts to apply concepts and practices in physical geography to an audience beyond the academic.

Indeed, the applied work of Brunsden, Cooke, Doornkamp and Goudie within the UK in the 1970s and 1980s in geomorphology could be viewed as an attempt to develop an applied niche for geomorphological analysis within civil engineering projects and environmental management in general.

By playing to this wider audience, it could be argued that these publications are of more significance in promoting the working practices of physical geography than are the textbooks mentioned above. Similar arguments could be made for US counterparts in applied work such as Graf and Schumm, just to name two. In this manner these workers were applying practical geographic knowledge in much the same way as their nineteenth-century counterparts.

In addition, the significance of graduates pursuing a non-academic career has rarely been considered in subject histories. Graduates who take up posts in consultancy or in management are in positions to influence greatly the type of physical geography applied to 'real' world problems. The trajectory of such individuals, and their use of the academic tools they learnt, has not really been discussed within the literature of physical geography.

This may partly be due to the perception of physical geography as a non- vocation subject. The potential impact of these key figures in promoting and implementing physical geography practices could have aided, via feedback to academics, a more policy orientated approach to the subject with all the baggage of moral and ethical dilemmas that this entails.

The development of a process orientation to physical geography meant that the definition of the discipline itself began to change. The need to reach into other disciplines for the necessary theoretical, field and increasingly mathematical techniques for analysis grew as each subdiscipline became more acquainted with the new focus.

Gregory argues that many individuals became familiar with the new focus through specific textbooks associated with each subdiscipline; for example, desert geomorphology via Cooke and Warren (1973); Embleton and King (1975a, 1975b) for periglacial and glacial landforms; and Carson and Kirkby (1972) for hillslope forms.

This meant that the core of study of physical geography became harder to define. The quest for an integrative explanatory framework for physical geography became somewhat superseded by a search for disciplinary homes with their own, existing frameworks. Whereas under the cycle of erosion the reconstruction of the environment was the core of study once process became the focus of study, each subdiscipline became concerned with process agents.

Studying these could involve the analysis of mathematical models of slopes, or the detailed physics of wind motion to the habits of limpets in weathering rocks. The common focus of study disappeared in the mêlée of process agents. Although it could be argued that the spatial variation of these agents gave each study its geographical flavour, it could equally be argued that the geographer only added an additional viewpoint to an existing field. This problem of discipline definition is a matter that physical geography still has not been able to answer.

Important to the development of a process basis for physical geography was the development of techniques for quantifying landforms and the landscape. The trend to increasingly complex representations of the physical environment within each subdiscipline further added to the separation of the subject. Within climatology and ecology, this trend for making description numerical had long been under way. Climatology had established its data-gathering credentials in the nineteenth century and had encountered and dealt with, albeit in its own way, problems of representation and scale, long before geomorphologists began to grapple with these issues. Likewise, ecology had developed techniques to identify and even simply count the entities of interest at a range of scales by the time geomorphologists got around to considering the problem of what to measure. Quantification of the landscape, or rather components of the landscape, was essential to enable assessment of the contribution of specific processes to the operation of the landscape system.

Despite the great advances made in technology, data collection and a re-orientation towards process studies, physical geography in this period still retained the roots of its past. Despite the new theoretical framework of plate tectonics, physical geography, or rather geomorphology, still retained a focus on landscape development, but at a more human scale.

This promoted relevant studies of landforms and landforming processes within such features as landslides and, more generally, in environment management. Despite the greater volume of information, the push for relevancy and the focus on process studies, the basic tenets of physical geography remained the same. The search for universality, the emphasis on the empirical and a concern with change, albeit as equilibrium and process-response rather than as Davisian stage, remained at the focus of work in physical geography.

The perennial problem of scale remained as a sticking point for the integration of processes found at different scales and acted as a brake on a purely reductionist view of the scientific endeavour in physical geography. A further trend that has been identified by several authors (e.g. Frodeman, 1996; Baker, 1999; Rhoads and Thorn, 1996c; Lane and Richards, 1997; Lane, 2001) is an increasing recognition of the importance of philosophy for the practice of physical geography. Although this may not be a surprise, given the purpose of this book, the manner in which this concern is reflected in the work of physical geographers should not be expected to be consistent. The new concern with philosophy does not reflect a new search for a universal framework for explanation, but rather a questioning of the ability of physical geographers to reach such a consensus.

One outcome of this new concern has been the recognition of the importance of studying small areas in great detail; for example, the significance of small-scale studies as noted by Lane (2001) in relation to fluvial geomorphology. Focusing at this scale of local detail, the universal and contingent become intertwined. Explanation and prediction become more difficult, but also more interesting, as the application of a general solution is not possible without the 'local' also entering into the equation.

SUMMARY

Ideas in physical geography differ over time and space. Development of understanding about reality can follow one of three models: progressive, paradigms or research programmes. A progressively better and truer representation of the world is achieved by building upon the successes of past researchers in the first model. Knowledge is cumulative and towards an absolute understanding of reality as it really is. Paradigms focus a group of researchers about a specific set of ideas about reality. Paradigm choice is as much about social pressures as about logical, objective choice. There can never be certainty that the understanding derived from one paradigm provides a true representation of reality. Confidence that specific paradigms do provide a true representation of reality may be high, but they can never be absolute. Research programmes highlight that researchers do cluster around specific ideas, but their choice of these ideas is based on logical and objective grounds of theory success.

The status of these programmes as true representations of reality is problematic and unlikely to be absolute. Although it is not possible to write the definitive history of physical geography, it is possible to pick out trends that seem to permeate much of the history of the study of the physical environment. Three key themes are identified: the search for universality in explanation, the study of stability and change in the physical environment, and the primacy of empirical information to study the environment. The three themes are related and have evolved together, each affecting and encouraging changes in the others. From these themes have developed the notions of a uniformitarian-based approach to processes in the physical environment. Likewise, the rise of Darwinian evolution and evolutionary interpretations of change in the physical environment have developed from the study of contemporary forms and processes. Central to supporting these ideas was the collection and classification of empirical data about natural phenomena. These themes survived into the late twentieth century, but the questions asked altered. Process-based studies and a concern with complexity echo the themes of universality and change, but with increasingly complicated empirical data to analyse the physical environment.

2

THE NATURE OF REALITY

What is reality?

Before analysing how physical geographers understand and measure the world it is important to discuss the nature of the reality physical geographers believe they study. This may seem like an odd place to start, as it seems clear that the physical environment is a real and solid thing that is open to study by our senses and the various instruments we have designed to enhance our senses. Such an initial assumption immediately sets up the concept of an objective, real reality of which physical geographers can have knowledge. Indeed, improvements in methods and theories to explain this reality could eventually result in a complete understanding of the physical environment. With this view it is easy to fall into the trap of believing that what is sensed and measured is real; that there is a direct and absolute correspondence between what we think exists and what really exists. All our models of reality capture at least part of a true representation of reality. Refining our measurements or theories, or both, results in a 'truer' representation of reality. In this view of reality, scientific progress means increasing our knowledge towards a true representation of reality, an absolute understanding of it.

The correspondence view of reality outlined above is difficult to sustain in the light of wholesale changes in concepts in areas such as geology, geomorphology and ecology. The introduction of concepts such as plate tectonics, complexity theory and non-selective evolution all imply that previous views of reality were wrong at worst or incomplete at best. How can we be sure that these 'new' concepts are true representations of reality and that the old ones were wrong? The old concepts were thought to be true representations of reality, so their replacement implies that there is something wrong with the correspondence view. A more appropriate view of reality may be that our current theories provide us with a coherent and non-contradictory representation of reality. In this coherence view, theories seem to work, they seem to explain what is sensed and measured, but there is no certainty that they are true representations of reality; theories can be subject to change. This means that theories tend to be coherent with our current state of sensing the physical environment, and with our current ideas about how the physical environment works.

An important point, however, is how do you select which theories are the most coherent? What criteria do you use and who has the final say? The arguments become very similar to those for and against paradigms and research programmes. The scientific community seems to develop and agree theories, but this is not necessarily a guide as to how true those theories are. Despite the seemingly conditional nature of theories, there is an implicit assumption that some contemporary theories are truer representations of reality than older theories. Science improves and progresses, our theories begin to mirror reality in a clearer fashion. Even if we recognize that we have not got an absolute understanding of reality, we feel we have a better grasp of it than we did in the past. Unfortunately, there is nothing in a coherence view of reality to support this assertion. All we can say is that our current theories seem to cohere with what we sense; we still do not have any absolute basis for believing our theories are better, or closer representations of reality, than theories in the past.

Being extremely pessimistic we could just plump for a more utilitarian definition of reality. Our theories make predictions that happen, our theories produce useful results so they seem to be useful: the pragmatic utility view of reality. This view tends to be seen as the least acceptable by scientists. It makes no claims about any movement towards an ultimate knowledge of reality. It makes no claims that our theories are more coherent than theories of the past, nor that coherence with current beliefs is the most appropriate way to assess the worth of a theory. Instead, usefulness replaces the quest for truth as the basis for theory selection and representation of reality. Taking a pragmatic view is not, however, necessarily a denial of a consistent working method for scientific investigation. Rescher (2001), for example, outlines functionalistic pragmatism as a possible approach to scientific investigation. Theory is central to this view of scientific investigation. The success of a theory is judged on how successful it is in its practical implementation. Success, of an instrument, or methodology, or procedure, lies in its successful application. Functionalistic pragmatism is not concerned with the truth of reality, but with how truths about reality are thought to be identified, with the processes of scientific endorsement. It is concerned with how and why 'truths' are validated. Success is defined by how well practices provide answers to our particular goals. Importantly, however, Rescher makes the point that although goals and purposes are important and can be matters of taste, evaluation is not. Evaluation requires a rationale, it requires rules and procedures that permit comparison of 'truths' between individuals.

The above views do share a common foundation by accepting that there is an external reality that is capable of study. We may not be able to get a real or true representation of that reality, but there is definitely something out there to study and, more specifically, to interact with. In addition, there is an implicit assumption that the reality we study has some causal basis. There is a reason, which is discoverable, for why something happens. Events in the physical environment are determined by other events and mechanisms; it is a deterministic universe. Such a view stands in distinct contrast to some of the philosophical stances of human geography, such as humanistic views of reality. The idealist view of reality as constructed in our own heads, the phenomenological view of reality as about deriving essences, and the essentialist pessimistic view of reality as angst and woe do not assume that there is something external to humans. The assumptions of being able to understand reality without reference to some external and real environment are difficult to find counterparts for in physical geography. Although some have suggested that a quantum-mechanics-based view of physical geography implies an idealist view of reality (Harrison and Dunham, 1998), reality does not exist until we measure it or interact with it; there have been few, if any, attempts to implement such views.

Assuming an external reality does not necessarily mean that scientists have to assume they will eventually understand that reality in full. Likewise, this assumption does not mean that scientists need assume that their representations are of reality as it really is. What the assumption of an external reality does provide is a basis for intersubjective communication – a basis for discussion within a common framework. One scientist may believe they have identified a process or an entity. Their analysis will probably be based on commonly agreed practices and involve criteria for identifying and manipulating the process or entity. This means that other scientists have a basis for recreating and redefining the process or entity identified by any single scientist.

Hall (2004), for example, used data from short-term measurements of temperature variations in a stone block to identify the release of latent heat when water freezes in the stone. He used this information to suggest that standard views of freeze–thaw were inappropriate as they overestimated the frequency of this weathering process. The proposition is supported with evidence from sensed data that is available to other scientists to reinterpret, Hall noting that one reviewer suggested that instrument sensitivity or reliability might explain the slight change in temperature measured. Similarly, Hall describes the monitoring set-up in sufficient detail for other scientists to copy his whole monitoring procedures to assess if they get similar results. Hall's assertion is, in other words, assessable by other scientists because they all assume that what he is observing is some aspect of an independent reality that is open to interpretation both in how it is sensed and in the replication of his observations. Without assuming that there is a common reality, such refinement or even redefinition of phenomena would not be possible.

VIEWS OF DIFFERENT PHILOSOPHIES

Despite the assumption of an external reality there is no single philosophy covering all of physical geography. This problem is made more acute by some attempts to graft various philosophies on to what physical geographers do. Reviewing, with the benefit of hindsight, 'old' masters or even current practitioners of physical geography as really practising a contemporary philosophy without knowing it is fraught with the same problems as developing Whiggish histories of the subject. It is important, however, to be clear about the different philosophical standpoints that have been advocated in physical geography and to be aware of their similarities and differences. Central to identifying the nature of the four philosophical standpoints considered here – logical positivism, critical rationalism, critical realism and pragmatic realism – are the ontology and epistemology of each. Ontology refers to what each philosophy believes is real. Raper (2000) identifies ontology as being concerned with concepts of identity; the way in which reality is identified and ordered. Epistemology refers to how each philosophy believes we can know that what we think is real is real. It is the study of how knowledge can be acquired, how it can be validated. Together these provide the basis for a detailed consideration of each philosophy.

Logical positivism is usually held up as a relatively naïve philosophy with little to recommend it in contemporary physical geography. Its ontology is that what is real are the laws and entities that interact in reality. Reality is as it is because there really do exist the laws we believe to exist, and these underlie the operation of all entities that also really exist. We can know about this reality by observing or sensing the interactions of entities through the operation of laws. This is achieved by observing and, where possible, measuring changes or the lack of change in entities under the action of specific laws. From such an analysis we can derive the structure and nature of reality as it really is. Logical positivism has its roots in the Vienna circle of philosophies in the 1920s, although as a view of how science works it goes back further, depending on how far you want to push the argument that early scientists worked with a philosophy as coherent and consistent as outlined by the Vienna circle.

Central to the logical positivist is the idea that you can identify, observe and measure real entities and their interactions. From these observations it is possible to derive laws of behaviour that can be proven to be real by further observation. Critical rationalism, developed mainly by Karl Popper (1968) as an alternative to logical positivism, has the same basic ontology and epistemology. Critical rationalists believe that there are entities capable of observation that interact in a regular manner. The key difference between the two philosophies lies in the belief of the possibility of obtaining truths about reality. Logical positivists believe that it is possible to assess a theory and determine if it is a true reflection of reality. We believe something because we can prove it.

Popper's critical rationalism denies that it is ever possible to be certain that a theory really does reflect reality. All we can do for certain, according to Popper, is to show that a theory does not reflect reality. Theories can be disproved, falsified, but they can never be proved. Falsification of a theory becomes the hallmark of the difference between science and non-science. This view does, however, tie Popper to a continual conditional view of reality. There can never be absolute certainty that our theories match reality. There is no gold standard of truth to which to compare our ideas. This could be seen as a problem as it implies that even cherished and seemingly correct theories and their outcomes, such as gravitation and the rising of the sun, are not absolute certainties. Popper suggested that although theories could not be proven there could be increasing certainty of agreement with reality, increased verisimilitude or truth-value – but never really stated how this worked in practice.

At the heart of both these views of reality is the use of induction and deduction in developing scientific explanation. Inductive reasoning believes that the 'facts' can speak for themselves. This could be seen as the 'vacuum-cleaner' approach to science. Collect as many facts as you can, and theories will spontaneously form and emerge full-grown into the world from this mire of data. Along a similar vein, inductive reasoning assumes that the truth of a statement is proportional to how often it has been shown to be correct. Statements become true by collecting data that show they are true. This form of reasoning puts a premium on data collection as the source and basis of theory and truth. Unfortunately, it is a logically flawed approach. Rarely, if ever, are data collected without a reason, without some guiding concept, that determines what type of information should be collected and how. Likewise, if a single counter-example to a theory is collected this destroys the logic of inductive reasoning. A single data point that does not fit the theory developed reduces the truth-value of that theory, breaking its supposed explanation of other data.

Deductive reasoning is based on the development of a logically, internally consistent argument. This approach has been called syllogistic reasoning: a conclusion follows logically and inexplicably from two premises. In its guise as the covering law model, the two premises are a set of initial conditions and a covering law. Applying the covering law to the initial conditions, the conclusion is an inevitable outcome. This form of reasoning provides consistent and logical statements. The only problem is that these statements need not bear any relationship to reality. Popper, however, used this structure of reasoning in his philosophy of critical rationalism. The derived statement could be viewed as a prediction of how reality should be if the covering law operates. It provides a statement that could be compared to reality to see if the covering law works under those initial conditions. Popper viewed the derived statement as a hypothesis capable of testing.

Unlike the logical positivists, however, Popper was intent not to prove the statement true but to devise ways of disproving or falsifying it. Falsification became the key criteria for judging if a theory was scientific or not. If no statement could be derived that was testable, then the theory was non-scientific. A standard comment at this point is that this meant that Einstein's theory of relativity was defined as non-science until 1918 when its predictions could be tested by a solar eclipse. This is not quite the case, as the theory of relativity was capable of falsification. There was an experimental situation that could be envisaged with the existing technology in which the theory could be tested, in which it could be capable of falsification. The specific circumstances required to undertake the crucial experiment had not been available until the solar eclipse of 1918. This makes the rigid demarcation between science and non-science much hazier, as 'capable of falsification' is a more difficult idea to pin down than actual tests of falsification.

Falsification also implied that there is never any certainty about the theory you are trying to falsify. Popper's ideal was to test competing theories until one could not be falsified. This was Popper's method of multiple working hypotheses. Different theories generate different covering laws that when applied to the same initial conditions should result in different, testable hypotheses. Each hypothesis should be capable of being falsified by critical testing. Ideally, you should be able to eliminate every hypothesis bar one by such critical testing. At this point you accept the hypothesis that has survived falsification, and its associated theory, as provisionally correct. The theory was never accepted as true, never accepted as proved, just not capable of falsification at this point in time. This means that any theory is unprovable. Achieving absolute truth, getting to reality as it is, is not possible in Popper's scheme. Popper suggested that although no theory could be proved, as a theory withstood more tests to falsify it, confidence in that theory grew. The truth-value of a theory increased the more testing it survived, but it never became true. This rather unsatisfactory end point for a theory implied that key scientific ideas such as the law of gravity were not necessarily true, just unfalsified. Their stability as cornerstones of scientific thought came from the confidence derived from their continued non-falsification, and the lapse in interest in trying to falsify them resulted from this confidence.

Recently critical realism (Bhaskar, 1978, 1989; Collier, 1994) has been put forward as another possible philosophical framework for physical geography. Although there are many flavours of critical realism, there are certain important differences in their ontology and epistemology compared to critical rationalism. Critical realism views reality as both stratified and differentiated. Reality is made up of entities that interact according to underlying laws related to underlying structures; it is differentiated. This differentiation reality is also stratified. Entities at the level of physical and chemical particles are subject to laws at that level. Organisms, being composed of molecules, are subject to the same laws, but in addition their interaction produces new laws found only at that level of reality. These new laws are constrained by the laws at lower levels.

No organism can behave in a manner that would break or contradict physical and chemical laws. Within these constraints new relationships emerge which, although inherent within the lower levels, are not predictable from them. This means that laws at one level or strata of reality cannot be reduced to laws of another level. Evolution, for example, is not reducible to purely genetics, although genetics plays a role. Evolution occurs at the level of the organism and as such it is relations at this level that determine the nature and regularities of evolution. This also means that causation and explanation need to be appropriate to the level of the entities of interest.

A differentiated and stratified reality means that the level of our observations is not the only level at which explanation can or should occur. All reality is encompassed at the level of mechanisms and structures. These structures and mechanisms provide the range of possible relations that could produce events. Below this level is the level of the actual. These are the events actually produced by the interaction of structures and mechanisms. Below this level is the level of experience. These are the events that we can identify and sense. As you go down from the level of structures to experience there is a decrease in the events available for study. At the level of structures, there are a vast number of possible events that could happen. Structures do not exist in isolation. Interactions between structures eliminate some events, whilst permitting and even generating others. These interactions restrict the types of behaviour possible and so reduce the potential events that structures could generate individually. This contingent and

CASE STUDY

Critical Rationalism – an Example from Environmental Reconstruction

Battarbee et al. (1985) discuss the possible causes of lake acidification in a specific region of southern Scotland. They identified four possible causes of lake acidification for this part of southern Scotland: long-term natural acidification, afforestation, heathland regeneration and acid (humanly produced) precipitation. For each possible cause a plausible scenario could be developed linking effect to cause. Long-term natural acidification had already been identified in other temperate environments and may have been related to adjustment of the environment to the end of the last glacial. In this scenario acidification of lakes is likely to have increased little since 1800. Afforestation would alter the soil characteristics and increase the acidity of runoff to lakes. Heathland regeneration, specifically the increased dominance of Calluna vulgaris in Galloway in the wake of a decline in grazing, would again alter soil characteristics and so increase the acidity of the water reaching the lakes. Finally, acid rain from industrial and other activities would increase lake acidity by direct input from precipitation.

These four causes could be tested by analysis of lake sediment from Loch Enoch by looking at changes in diatom populations, pollen composition and heavy metals content. Diatoms are microscopic planktonic creatures with a silica skeleton. They live as communities in lakes and when dead can settle on the lake bed and become incorporated into the sediment. The composition of a population is highly sensitive to the pH of the lake water. Identifying the species present in a lake sediment permits a characterization of the overall structure of the diatom community and therefore indicates the pH level of the lake water. Pollen grains, at least those that enter the lake and are preserved in the sediment, indicate the nature of the plant community within the lake catchment.

Heavy metals are not found in significant quantities within the lake catchment. Any heavy metals within the lake sediments are therefore likely to have been transported into the catchment, probably via precipitation. Their presence points to pollution sources external to the lake catchment. Each of the above causes would be expected to alter the nature of either the diatom or pollen communities or the amount of heavy metals. This means that each cause can be rephrased as a simple question, a hypothesis, to which there is a simple yes or no answer; a falsifiable question. The four causes need to be rephrased as hypotheses and the criteria for falsification clearly identified. Long-term lake acidification can be restated as: Do diatom communities indicate that acidification has been either increasing steadily or has not increased in the last 200 years?

This question takes the data from diatom analysis as the basis for falsification. The question defines a specific time period to which the assessment should be applied and identifies the two conditions that would mean this cause could not be falsified. Diatom analysis suggests that acidification began in 1840, so the above question can be answered as no to both parts.

Afforestation as a cause can be rephrased as: Does the recent (last 50 years or so) afforestation increase the acidity of lakes in this region as indicated by diatom communities and by evidence from other lakes in the region? This hypothesis is assessed using not just the sediment core data, but also using information from other studies. The investigators reject this hypothesis by referring to the acidification of non-afforested catchments, such as Loch Enoch, as well as the evidence of acidification of other lakes beginning in the late nineteenth and early twentieth centuries, before afforestation of the lake catchments. The land-use hypothesis is assessed using the pollen in the sediment core in combination with the evidence of the timing of acidification provided by the diatom community data. The hypothesis can be rephrased as: Does an increase in heathland area as indicated by the plant communities found in the pollen analysis result in an increase in the level of acidification of the lake as indicated by the diatom communities?

The pollen data suggest that the plant communities have been relatively stable over the last 200–300 years and, if anything, that Calluna vulgaris has declined in the recent past, not increased. Finally, the investigators test humanly produced acid rain as the cause of lake acidification. The cause can be rephrased as: Do heavy metal fluxes in the sediment core increase as lake acidification as indicated by the increase in diatom communities? The question requires the identification of fluxes of heavy metals, specifically zinc, copper and lead, all metals that would not be expected to have high natural concentration in this type of catchment. The only source for these metals that the investigators entertain is acid precipitation from human sources. Background concentrations of these metals from the catchment show no abnormalities. The variation and increase in the concentration of these heavy metals in the sediment core can be correlated with the acidification of the lake as indicated by the diatom communities. This means that the question can be answered as yes, and so the hypothesis is not falsified. The conclusion of the paper is worth quoting in full, however: We cannot prove that an increase in acid deposition was responsible for the acidification of Loch Enoch and similar lakes in Galloway, but we have shown that alternative hypotheses, as presently formulated, are inadequate.

Nothing has actually been proven as true. The inability to falsify the last hypothesis is viewed only as indicative of what is likely to be the 'real' cause. There is the potential for other evidence and, as importantly, other formulations of hypotheses that could be assessed and falsified. Acceptance of acid precipitation as the cause of lake acidification is only provisional in the absence of other candidates. The conclusion is scientifically accurate in Popperian terms. It does not extend its results beyond the Galloway region to an area of similar geology, ecology or remoteness as the study site. unpredictable juxtaposition of structures produces events that actually happen. These events further define and constrain the future interactions of structures and so the actual can influence the development of the structural level. The level of the actual is not, however, the level that is studied. The level of experience is a much smaller range of events that we can identify and sense.

Positivism and critical rationalism tend to reduce reality and explanation to this final, empirical level. At this empirical or experiential level, the pure operation of structures is not likely to be clear. The power of individual, underlying structures to generate events has been influenced, potentially even nullified, by the interactions with other structures. All that is likely to be observed is a weak signal of the potential behaviour generated by an underlying structure. Pragmatic realism and semiotic approaches to physical geography have recently been suggested as an appropriate framework for understanding in physical geography. Baker (1996a, 1996b, 1999) looks back to the founders of geomorphology in North America and the practically minded philosophy of Charles Peirce (1957), semiotics, as a basis for developing a pragmatistic approach. Pragmatic realism is not a simplistic philosophy about usefulness being the measure of worth – this is the pragmatic utility view of reality. Ontologically pragmatic realism believes that an external reality exists with entities that interact with each other.

Epistemologically, as researchers, however, we can never be certain we have divided up reality as it is. All we know is that we have divided it up in a way that seems reasonable and consistent for our purposes – it is pragmatic in this sense. There may, however, be a number of ways in which reality could be divided and studied that are consistent and pragmatic; there could be a number of ways of world- making. Pragmatic realists accept that this is the case and that their models of reality are conditional and subject to debate and change. Having accepted this, pragmatic realists still believe that reality is underlaid by structures that produce law-like behaviour. We can theorize about these structures and laws, but as with entities we can never be certain that we have identified reality as it really is. Understanding itself develops by relational thinking. Understanding is only possible because we relate one entity to another; we only understand an entity by its relationship to other entities.

These entities are only understood, in turn, by their relations to other entities, and so on. Baker (1999) believed that physical geography should use the concept of signs to study reality. Baker suggests that, from Peirce, signs involve certain conditions. First, the sign must represent an object in some manner. This representation is based or grounded in the sign itself. The sign provides a quality to the object that exists within the sign. The example Baker uses is of a rock being observed as foliated.

The quality of foliation is a quality given to the object, the rock, by the sign. Second, the sign must be interpreted by an interpretant. A sign represents an object only if there is an interpretant to correlate or draw the two together. There is no base to this interpretative process, however. The foliated rock, for example, is not just foliated, it is composed of a past history and a current state that are also conveyed in the sign of foliation. The relationship between foliation and the rock texture it expresses is a further sign relationship, a relationship requiring an interpretant.

Baker makes the point that this chain of sign interpretants is never ending. The fossil in the figure has a relationship to its object, the past organism that it represents. Baker identifies this as a causal or indexical relationship (Baker, 1999, p. 641). Fossilization itself produces a kind of interpretant, the fossil within its geological bed. It is sign waiting for realization. Baker suggests that this sign can be triggered by a trained palaeontologist who can interpret its causal relations. The causal connection between the fossil and its past organism is translated into a mental process of connections in the observer. In turn these become further triggers for products, such as scientific papers, and further inquiry, all interpretable as signs in themselves. The important point of Baker's analysis is that the fossil on its own has no value or worth. It is only when it becomes enmeshed within a network of signs that it can be interpreted and act as a stimulus for further sign generation.

Significantly, this view of reality links the object and subject directly and blurs the distinction between the two. This is a theme that will be developed further in relation to entities. It is sufficient to note this now and suggest that this makes a clear objective/subjective division of reality almost impossible within pragmatics. This view of a continual shift between the objective and subjective in working practice in the study of the physical environment is not new. Van Bemmelen (1961), for example, noted that the interrogator of reality, the geologist, needs constantly to move from induction to deduction and back again in pursuing a comparative analysis of observation and evidence.

Reality as a Dialogue

The philosophies already outlined provide a snapshot of current thinking in science, but they also reflect a general bias in most discussion of scientific philosophies. The philosophies tend to focus on the 'hard' sciences, and in particular on physics as their model of what science is and how it should be done. This means that most discussions concerning philosophy may appear to be dealing in the abstract, but often use concepts of direct relevance to the practice of physics. Physical geographers have tended to assume that their approach to reality has to match that of the 'hard' sciences. Problems unique to trying to undertake what is variously described as a field science, a historical science or an environmental science have been viewed as merely local problems that should not concern the general philosophies used. This has resulted in post hoc justification for research designs, and particularly phases of reductionist work in physical geography, as essential to ensure that physical geography remains a 'proper' science.

There may be nothing inherently wrong with this view, but if it is taken as the only view of physical geography, the only way to approach this particular and unique scientific endeavour, then it can become highly restrictive. This view tends to perpetuate the idea that there is an objective reality from which we draw real knowledge. Reality is there, we are clever and extract real data from it to find out what reality is really like. This view is one that it is difficult to sustain once you start the practice of actually doing physical geography. At this point, another view of how reality is understood may become more appropriate: the view of reality as constructed by a dialogue. The metaphor that an understanding of reality is at heart based on a dialogue between scientist and reality is not a new one, and has been associated with physical geography as a historic science. Demeritt and Dyer (2002) identified this trend in both human and physical geography.

They identify different uses of the term applied to the research strategies of geographers, including literal and metaphorical. The metaphor of dialogue begins to make problematic what appeared to be natural. The metaphor highlights the importance of both the observer and the observed and continually switches between them. The use of metaphors, such as a conversation with nature or reading nature as a book, is common both as a teaching method and as a view about how to study nature (e.g. Cloos, 1949; Wright, 1958; van Bemmelen, 1961). Despite the common use of these metaphors, they have been little used to form a basis for trying to devise a philosophy of physical geography and its practice. Although these metaphors can be interpreted as backing up a very reductionist view of reality, implying that there is a single, correct reading of the book of nature for example, they can also be interpreted as implying a less rigid view of reality. Van Bemmelen (1961) noted, for example, that the character of the subject, the researcher, can become as important as that of the object of study. Indeed he compared the geologist to a doctor highlighting the 'art' of investigation as much as the 'science'. The dialogue metaphor strikes a particular resonance with the critical-realist and pragmatic-realist views of reality. The dialogue metaphor highlights the negotiated nature of reality. Reality is not just a thing to be probed and made to give up pre-existing secrets. Reality enters into the dialogue by answering in particular ways and by guiding the types of questions asked. Central to this metaphor is the interpretative nature of inquiry. This does not deny that an independent reality exists, only that it is the engagement with that reality which is the focus of study. Some pragmatic realists have argued that the assumption of a real, external reality is vital for any dialogue to work. Although absolute knowledge of the external reality can never be achieved, the idea that one exists means that researchers have an assumed common basis for reference and discussion of their understanding.

Each researcher is willing to enter into this dialogue because each recognizes the potential fallibility of their own view of reality, and so the potential for modification. Reality itself plays a constraining role for this debate amongst researchers, guiding research questions and methods, but not along a predetermined or predefinable path. The dialogue rather than the subject of the dialogue is what we can know about. This puts engagement with reality, or rather physical geography as it is really practised, at the heart of any philosophical discussion. This also means that the false distinction between 'theoretical' and 'practical' work dissolves, as practice involves both undivided. Similarly, the dialogue metaphor pushes debate away from other polar opposites such as 'realism' and 'idealism' or 'objectivity' and 'subjectivity'.

Rather than looking at these as 'either/or' debates, practice or the 'doing' of physical geography involves the investigator in continually shifting from the 'objective' of the entity under study to the 'subjective' of mental categories and concepts used to understand the entity. Continual shifting and informing of practice by the 'real' and the 'mental' means that neither can be understood in isolation. Practice brings out the relational nature of understanding in physical geography and the haziness of seemingly concrete concepts such as entities.

Theory, Reality and Practice

Defining theory precisely is highly problematic. Some authors provide rigid and highly formalized definitions of theory, whilst others question the value of the term at all. Theory can be viewed as a framework of ideas that guide what we think reality is and how to go about studying it. Abstract ideas such as force and resistance can be thought of as theoretical constructs. These two ideas can be linked together to suggest what should happen in reality. For example, if force is greater than resistance then change should occur. The two ideas form irreducible elements of the theory. They are the axiomatic elements. Their existence is not questioned; it is taken as given within the theory. Likewise, their interactions, usually based upon some set of physical principles or mechanisms, are the axiomatic principles of the theory.

The ideas are, however, very abstract and need to be translated or superimposed upon parts of reality. Theory needs to be linked to reality. This is usually achieved by bridging principles, rules that define how abstract ideas can be interpreted as real entities capable of identification and testing. Within science, these bridging principles are essential as the means by which theories can be tested and, in a pure Popperian sense, rejected or accepted.

It is interesting to note that Haines-Young and Petch (1986) distinguish between theories and myths. Within myths they include Schumm and Lichty's (1965) paper on time, space and causality, Wolman and Miller's (1960) paper on magnitude and frequency, as well as Hack's (1960) paper on dynamic equilibrium. For many geomorphologists in particular these papers would seem to be central to any understanding of the physical environment. Haines-Young and Petch, however, make the point that all these papers express what they term as 'truisms'.

For Schumm and Lichty, for example, they state that the truism is 'that theories explain variables and that only certain theories have been developed' (p. 139), a problem at the time of the paper, but, in their eyes, a trivial issue in retrospect. Wolman and Miller (1960) are merely stating the truism that whatever landscape feature is observed, it must have been formed by processes that had some magnitude and frequency (p. 139).

These papers, as well as Hack's (1960) outline of dynamic equilibrium, state ideas that cannot be disproved. They can be linked to reality in a multitude of ways excluding nothing, and can never be falsified because of this. Inability to focus or narrow down a set of ideas to form a distinct and singular set of bridging principles means that there is always scope for reinterpretation of the ideas in a favourable light; that is, the ideas are unfalsifiable. Myths are not useless in physical geography, however. Haines-Young and Petch point out that although these ideas are not theories, as defined within critical rationalism, and so are not sources of any understanding about reality, such myths form the basis of a lot of testable theories in science.

They view it as possible to derive testable and coherent theories from the myths. They see it as the task of a scientist to develop such links between myth and theory, to prompt questions based on these myths that are amenable to testing in reality.

Myths and Theories

The classic paper on 'Time, space and causality' by Schumm and Lichty (1965) is often viewed as a major turning point in thinking about geomorphology. Haines-Young and Petch (1986), however, suggest that the main argument of the paper amounts to a 'truism' or myth rather than a theory. The basic idea is that at different temporal scales, different variables will be independent, dependent or irrelevant for understanding the functioning of the landscape. The temporal scale of analysis you are interested in determines which of the variables you need to study. Undertaking a study of a catchment within graded time, for example, would require an analysis of relationship between dependent and independent variables. Variables such as time and initial relief are irrelevant for your study as these are unchanging over this time-scale.

Haines-Young and Petch (1986) characterize this point as merely the statement that geomorphologists have developed different theories to explain landform behaviour at different scales. The variables vary in their status as dependent, independent or irrelevant, because the theories used change. It is not the variables that alter their status, it is theories that alter to explain a different variable at a specific scale. Tinkler (1985) notes that the key point to Schumm and Lichty's analysis was to point out that theories actually relate to different things at different scales. Acceptance of one approach, the historical, is not incompatible with another, say that of equilibrium, as each deals with different things at different scales. Each is mediated by different theories that identify different objects of study.

The ideas of Schumm and Lichty are not falsifiable. They are only statements of how research into reality is structured within geomorphology. The idea that theories relate to different variables at different scales does not generate any hypotheses to assess. The idea could, however, provide a basis for directing thinking about reality. If theories relate to different variables at different scales, then an investigator is not bound by an overarching single scale-independent theory. What is an appropriate theory, an appropriate approach to study, could vary with the scale of object being studied.

Indeed, the objects themselves could change as you change scale. It is in this sense that the 'myth' of scale put forward by Schumm and Lichty is of use in geomorphology. Line is between the myth and the derived theory is unclear. There may be many ways of undertaking such a translation from myth to theory, and so the problem of an unfalsifiable base remains, despite the presence of a falsifiable product, a theory. Von Engelhardt and Zimmerman (1988) suggest that theories are hierarchically ordered systems of hypotheses interlocked by a network of deductive relationships.

This highlights the significance of hypotheses to one view of theory and possibly may give myths a role as a higher-level structure. They view theory, hypotheses and the empirical levels of reality as interlinked. Using the theory of plate tectonics they construct a hierarchical framework of theories, hypotheses and 'facts'.The highest level of the theory are the basic hypotheses from which the partial hypotheses are derived.

From the partial hypotheses are derived the inductive generalizations and empirical basis of the theory. Each level informs the others. Note, however, that the basic hypotheses are few but that the hypotheses derived from them are many. Likewise, the empirical 'facts' are more in number than the partial hypotheses that they inform.

The structure of an increasing number of derived statements is a common one. Both empirical and theoretical 'facts' (by which they mean theoretical concepts such as the atom) can both be used to construct theories. Within physical geography it can be assumed that most theories will contain both and be informed by the success or failure to explain or predict the behaviour of these entities. Von Engelhardt and Zimmerman draw little distinction between theory and hypotheses regarding the two as related to each other.

Theories only differ in being further up the hierarchy and being greater in the scope of their application, their significance as explanatory structures, and being more reliable in terms of consistency. According to von Engelhardt and Zimmerman (1988), a theory should be both internally consistent, in terms of its logic structure, and externally consistent, in terms of its consistency with the hypotheses derived from other theories. These other theories include the underlying physical and chemical theories that usually make up unquestioned components of theories in physical geography. They do recognize that both theory and hypothesis represent generalization, abstractions of reality. They are not mirror images of reality, they describe and explain only the empirical and hypothetical components of reality and so are only indirectly reflecting reality as it is.

Theories can be identified at a number of levels in any study. A theory about force and resistance within fluvial systems requires bridging principles to link its elements to measurable entities within the fluvial system. If force is equated with flow velocity then measurement of flow velocity becomes a surrogate for this part of the theory. But how can flow velocity be identified and measured? Flow velocity will vary across the river profile, so there need to be criteria for selecting precisely where flow is measured.

Watching a river you may notice that flow need not always be directly downstream. River flows swirl and are deflected, forming eddy currents of various sizes. Again agreement is required concerning the appropriate point or section of river to measure flow. The instruments used to measure flow can vary from an orange and a stop-watch along a measured reach to an acoustic or a laser Doppler device. How can we be sure that these instruments are measuring the same thing – river flow? Instruments such as the laser Doppler-flow monitor measure movement based on specific physical principles about how reality works. In other words, our theory of river flow as a force is being measured by an instrument that is itself designed and developed on the basis of a theory about reality. If the principles upon which the instrument is based are wrong then the measurement of flow is flawed.

All observations are theory-laden to some degree or other, as Rhoads and Thorn (1996b) noted. The question is, does this theory-ladenness matter to the final identification and testing of the theory at the level of interest to the physical geographer? Identifying that laser Doppler devices are dependent upon a theory about how reality works developed within physics is something inherent to all instruments of this type. The theories associated with the production of laser light, the interaction of the laser with the environment, are all assumed to be correct. Without this assumption there could be no assessment of theories developed at the level of the entities of interest to physical geography. We would instead be continually questioning and testing the physics and chemistry of our methods.

Bogen and Woodward (1988) provide an interesting spin on theory-ladenness, a viewpoint developed by Basu (2003) in relation to the work of Priestley and Lavoisier on oxygen. They suggest that, contrary to popular belief, theories do not make predictions or explanations about facts. Instead they believe that a distinction has to be made between data and phenomena. Data have the task of providing evidence for the existence of a phenomenon and are relatively easy to observe. Data, however, cannot be predicted or explained that well by theory – data just are. Phenomena, on the other hand, are detected through the use of data, but, in all but the most exceptional cases, are not observable. Data stand as potential indicators of a phenomenon, but not necessarily just one phenomenon. Data could be used to identify many different phenomena depending on what the data are thought to detect. Phenomena are viewed as being generated by establishing a particular experimental design.

The design is produced to identify and monitor the phenomena. There are a limited number of circumstances that will permit the phenomena to emerge from the experimental situation. It would be expected, however, that phenomena would emerge once the limited circumstances had been obtained. Data involve such disparate and numerous combinations of factors, often unique to the experimental situation, that no general theory can be constructed to explain or predict data.

Phenomena have stable and repeatable characteristics that can be detected, provided the appropriate circumstances are present, by a variety of different types of data. The different types of data need to be produced by the phenomena in sufficient and appropriate quantities to enable their detection and interpretation within the theory being assessed.

Data cannot, therefore, be produced by any means, they need to be produced within a context appropriate for the assessment of a phenomenon. This means that data production is informed by the theory under assessment and will be restricted in its location and timing of production as dictated by the phenomenon under study. If a phenomenon can only be isolated in the laboratory, then only data produced within the experimental confines of the laboratory will be considered as relevant. Production of such data may involve long processes of establishing adequate control and techniques for ensuring data quality, as well as alteration of the data to a form amenable to appropriate statistical analysis.

Physical Geography as Historical Science

There has been a tendency, both historically and currently, to try to set the working practices of physical geography as a separate type of scientific explanation: as a historical science. Some practitioners have assumed the inferior status of a historical science. Andersson (1996), for example, in relation to explanation in historical biogeography, stated that the explanatory structure of mathematics, physics and chemistry was clearly the most prestigious model of explanation with the greatest explanatory power, but one hardly applicable to historical biogeography.

Johnson (1933) identifies seven stages to investigation in historical sciences. Within each stage he believes it is vital to undertake detailed analysis. By analysis he means detailed and thoughtful testing of the basis of each investigation at each stage or 'tracing back each part to its source and testing its validity, for the purpose of clarifying and perfecting knowledge'. His preferred method for studying the environment is by the use of multiple working hypotheses, through which different ideas can be tested and analysed.

The basis for distinguishing the practice and explanations developed in physical geography – whether it is in the guise of ecology or geology – from those of 'traditional' or experimental science, lies in the subject matter of study and the explanatory framework sought. A historical science is not necessarily concerned with generalities, instead it is concerned with explaining particular events, both present and past: Similarly, in geology we are largely interested in historical 'individuals' (this outcrop, the Western Interior Seaway, the lifespan of a species) and their specific life history. It is possible to identify general laws in geology that have explanatory power . . . but the weight of our interest lies elsewhere. In explaining specific individuals and events reference may be made to general laws derived from 'hard' sciences, but they will be used as part of the logical arguments concerning the specific individuals or events. The 'laws' will not be the explanation of the specific, instead they will form part of the explanatory framework.

The historical scientist has observed phenomena as their focus of study, but without the luxury of observed causes (Cleland, 2001). In experimental sciences, causes are observed, often under controlled conditions, and phenomena can be observed as produced by them under the controlled conditions.

Historical science has to deal with what Cleland (2001) calls the asymmetry of localized events. An event such as an eruption produces a multitude of effects of which only a few need to be identified to infer that an eruption has occurred. A single effect cannot, however, be used to infer a cause. There are many possible and plausible causes that can be linked to a single effect. Historical sciences have developed strategies – that of multiple working hypotheses being one – with which to whittle down the number of cause-and-effect relationships. Historical science searches for the most plausible set of cause-and-effect links as the explanation for a phenomenon.

Historical science looks for what Cleland calls the 'smoking gun', the piece of evidence that unequivocally links a specific cause to a specific effect. Failing a singular, decisive piece of evidence, historical science instead focuses on identifying a subset of effects, or 'traces' as Cleland calls them, which can be viewed as evidence that only a specific cause could have produced them.

Tradition and experience will identify the small subset of traces or types of evidence that usually provide information for this judgement. Within environmental reconstruction, for example, pollen analysis or stratigraphy is probably the most widely adopted and versatile method (Lowe and Walker, 1997). Pollen can be extracted from different sediments and used as evidence for a whole range of environmental conditions.

Causation within historical science is recognized as a complicated and potentially complex affair. Sets of traces, or evidence, are built up to construct a picture of the past and the manner in which past events caused effects in the past and resulted in the present state of affairs. It is often unclear where causation lies within this picture.

The bursting of a glacial dam might produce a catastrophic flood that alters the course of a river and is a major formative event in the development of a number of landforms. Identifying evidence of the dam-burst might be supposed to have identified the cause of the changing river course or of the slope profiles of the catchment. The dam-burst may have triggered the effects, but is it the cause? The flood of water would have had a lesser impact had the river channel been highly constrained by existing geology. Likewise, a large flat floodplain with few steep slopes would not be affected by the catastrophic flood in the same manner as a newly deglaciated landscape with unconsolidated moraines scattered across it.

In other words, the context within which the trigger event occurs is as important as the trigger itself in any explanation. Simpson (1963) noted that the immanent or ahistorical processes, that experimental science is good at identifying, need to be considered with the configurational aspects of a situation to understand or explain reality.

The configurational is the series of states that have uniquely occurred through the interaction of ahistorical processes with historical circumstance – the context. Without an understanding of both, causation may be mistakenly attributed to a single event. Of itself, the trigger is not sufficient to cause the effect; the trigger is only a cause because of the specific configuration of reality it occurs within.

It is worth noting that historical science also has a major problem with identifying the nature of the individuals it is trying to study (Hull, 1978, 1981). Individuals do not appear fully defined and fully formed for study.

A set of characteristics are needed to differentiate the individual from its context. Furthermore, the individual may alter in these characteristics over time, as its individuality is degraded and becomes increasingly undifferentiated from its context. For example, as a mass of material moves down a slope at what point is that mass no longer a part of the slope? When does it become an individual in its own right? Alternatively, as soil properties change, moving across a landscape, at what point does one soil type become another? Although there may be standardized classification criteria for identifying soil types, is a distinct value at which a soil is one thing and then another appropriate to understanding reality?

Summary

Physical geographers study a reality that they assume is external to themselves and that is capable of study. Although they may never have absolute knowledge of this external reality they can organize and use information to argue and come to an agreement about the nature of their representations of reality.

Logical positivism believes that it is possible to sense reality as it is and that it is possible to discover real laws about real entities through sensor information. Research is directed towards proving reality to be as we believe it to be. Critical rationalism takes a more critical view of our search for an understanding of reality. Critical rationalists view the testing and falsifying of ideas as the essential feature of scientific research. Knowledge is never certain or absolute, it is only the ideas which we cannot disprove that we accept as true for the time being. Within both philosophies deductive and inductive reasoning are used to reason about reality. Deduction is the only logically consistent method of argument, but unless it is connected to reality via empirical assessment it can produce invalid arguments.

Critical realism highlights the differentiated and stratified nature of reality, the asymmetry between the real and the actual. All we can observe is the actual, but this is a pale reflection of the structures and mechanisms that underlie reality. Assessment of ideas is more problematic in critical realism, but testing of hypotheses about reality is still a central means of understanding. Pragmatism emphasizes the importance of the mental image the researcher has of reality in understanding that reality. Any testing of ideas makes use of entities as signs, as signifiers of ideas and divisions of reality. The interpretative network of signs of which the entity under investigation is a part is not a closed network. Relations are always open to reinterpretation and entities can undergo renegotiation. Object and subject of investigation cannot be easily separated. Understanding reality increasingly becomes a dialogue between the socially embedded researcher and reality. The ideas of the researcher are guided by theory, as are the means by which reality is investigated and the methodologies employed. Theories provide a framework for deciding what to study, how to study it and how to interpret the outcome of the dialogue.

3

ENTITIES AND CLASSIFICATION

INTRODUCTION

Entities, which are vital in any study, are often viewed as unproblematic and distinct divisions of reality. Traditional practices, entrenched theories and increasingly advanced monitoring technologies can all conspire to make entities appear unproblematic. Entities form the basis for the units upon which theories, and their translations to assessable hypotheses, are based. The entities studied indicate how researchers believe reality operates; how it is divided up, and by what processes divisions are made. Despite their pivotal role, entities remain relatively poorly conceptualized. Problematizing the status of entities seems to many to bring into question the whole process of scientific study. Questioning entities, however, is a key part of the dialogue between researcher and reality. What is being studied is being continually redefined, renegotiated and refined. The whole process of research results in an often slow, occasionally rapid, invisible renegotiation of an entity. The same name may be used, but the connotations associated with an entity change.

It is clear that physical reality does not change as paradigms change. It is even likely that the property-based definitions of mountains do not alter. So what did change? What changes is how the entity or individual is defined and the classes into which these individuals could be divided. Therefore a central concern has to be what do we mean by change in entities? Are we discussing a physical reality or a mental construct? Unfortunately, the answer is both. This is because physical reality can only be grasped via our theories. If these alter, what we believe physical reality to be alters – even if it appears that the entities remain the same physical things. Entities become defined and entrenched within particular theories, which in turn provide the basis for identifying new properties, that further help to define the entity, as well as guiding how the entity should be measured. In other words, new theories renegotiate entities in their own image. Physical geographers have negotiated, although they would probably generally view this engagement as improving their understanding of reality rather than as a process of renegotiation. This chapter tries to address the issues above by looking at, first, the nature of entities, what they are, and how they are defined, as well as why they are important. Of importance in this discussion is the relationship between entities and kinds. Kinds are linked to the issues of classification and standardization of reality. The act of classifying and standardizing gives an impression of a fixed, unchanging and therefore objective reality.

What are Entities?

Philosophically this is a tricky question. Entities are the units we believe exist in reality. They are the things we try to study. How they behave, and why they behave in that manner, is what we are trying to explain. Whether they really exist as we think they do, however, is a more difficult issue to clarify. We may, for example, study a landslide and try to explain its behaviour by reference to smaller parts. By focusing on the component parts, does this mean that the landslide is not really the focus of study or the basis of explanation? Are not the component parts the focus of study and so the things that really exist and hence the 'real' entities of study? Could we go even further, and believe that only physics and chemistry study the real basic entities? Physical geographers merely study aggregates of these basic entities. The next section will highlight why, within physical geography, such a reductionist view of entities, and the reality they represent, is misplaced.

A key characteristic of an individual entity is that it is assumed to exemplify a distinctiveness through having a particular property or attribute or set of these attributes (Loux, 1998). However, viewing entities as isolated islands of distinctiveness is not of great use in explanation. Entities need to be viewed as signs or as representative of something else – as illustrations of universals, for example – to be of use in explanation. There are at least two possible views of entities (Loux, 1998). One view is that reality is made up of universals. These universals are real entities that find expression, at least partially, in entities we find in our 'real' or experiential world. Such universals can be exemplified simultaneously by different objects, as there need not be a one-to-one correspondence between universals and experiential entities. Universals define the properties a thing has, and its relationship to the kind to which it belongs. Mudslides have a set of properties by virtue of being mudslides.

These properties are universal, possessed by any entity defined as a mudslide. Mudslides we observe are not 'pure', however. The realization of the full set of properties may be polluted by the nature of the material the mudslide occurs in, the specific environment conditions it occurs under, and so on. We know that these properties are not fully expressed because we have in mind the ideal type, the universal mudslide, which exists in reality but only finds expression in the pale realizations we can observe.

Another view is that reality is only made up of particulars, of individual entities themselves. These entities are not examples of some underlying universals. Instead, each particular has properties that may or may not be the same as properties possessed by other particulars. In this view two things may occur that possess the same velocity of movement, that are composed of roughly the same material, and seem to behave in a similar manner as far as we can identify and measure their behaviour. They are both called mudslides. These mudslides have similar properties because we say they do. Possession of similar properties is not an indication of any relationship between the two entities, of any prediction of behaviour of one from the other. This possibility would imply that both entities are linked in some manner to a universal. Both views of reality do, however, accept that entities can be structured and complex in nature, and not the simple basic units of reality a more reductionist view might envisage. Properties that are the hallmark of a distinct entity can, however, be held in different ways. Any entity could be viewed as an entity shell devoid of any properties – the so-called substrata view. Properties are independent of the entity and not necessary for its existence.

An entity exists and properties fill it. Despite being associated with an entity, the properties do not define it. Almost diametrically opposed to this view is the concept of an entity as a bundle of properties. There is no empty shell to fill. Instead, it is the association of properties or attributes that define the entity. An entity has no physical existence outside of the potentially contingent association of properties. Without properties the entity melts away; it has no substance. Both these views have severe metaphysical problems, and within physical geography neither would be able to enhance explanation. The first assumes the existence of an entity without properties, the very things we identify, define and by which we measure an entity. The second provides no solidity to a form, just viewing an entity as, at the extreme, a contingent and potentially random association of properties. An understanding of reality based on either would be very limited.

Entities and Kinds

A way out of the explanatory dead-end of the above views of entities is provided by the concept of substance or essence. The idea has been traced back, as much of philosophy, to Aristotle. The central tenet of the idea is that entities can be seen as irreducible. Irreducible entities are not merely the sums of their parts, they are more than this. When put together the whole has properties grounded in universals that refer to the essential nature of the substance of the entity. It is these properties that define an entity. This means that not all properties associated with an entity are of equal importance for the essence, the substance of the entity. Without a limited subset of specific properties the entity could not exist or be identified. These properties define, or rather permit the researcher to identify, the substance of the entity. The properties are not, however, filling an empty shell.

The properties exist, take the values they take, and enter into the relations they do, because they are part of a specific entity. The relationship between entity and properties is mutual and indissolvable. A key universal is that of a kind. Kinds provide an identikit of the essence of an entity. Membership of a kind defines those properties that should be present in any entity of that kind. Non-essential properties do not alter an entity's membership of a kind, but they can mean that entities of the same kind need not all have properties in common, nor essential properties with the same values. Kinds can be thought of as templates, outlining the basic contours of what it means to be an example, a member of a kind. Kinds can also form nested hierarchies of more general or more specific essences, depending on which way you move through the hierarchy. Animals, for example, could be seen as a general kind of which apes are a more specific kind and humans an even more highly specific kind. Each more specific kind retains the general properties of the more general kind, but is differentiated by other properties which do not contradict or prohibit their membership of the more general kind. Linnaean classification in biology could be seen as a classic example of nested kinds.

Providing a template for the essence of an entity is fine, but does it aid explanation? If you had an entity of a certain kind you would expect it to behave in a certain way. Membership of a kind defines the relationships possible, the essential properties, and, potentially, how an entity can change. But is the kind then the appropriate focus of explanation? Could not a geomorphological feature, such as a slope, be composed of other smaller kinds such as soil? Are these smaller kinds more fundamental than the slope? Are they more appropriate as foci of explanation? Could you keep this reductionist regression going until you reach 'real' kinds, often referred to as natural kinds? Rhoads and Thorn (1996b) highlight the significance of the object rather than the method of study for geomorphology. They discuss whether landforms are natural or nominal kinds. Natural kinds they view as having some objective and real nature in reality, whilst nominal kinds are humanly constructed artefacts.

If natural kinds exist, their objective nature could imply a superior status in any explanation. Explanation involving natural kinds would be viewed as involving real things, reality as it is, rather than a potentially subjective and changeable human artefact. Studying an entity as a natural kind has distinct advantages. Researchers can assume that they are studying the world as it really is and so obtain objective, independent information. Investigation of natural kinds derives the essences of 'real' objective entities, their generative mechanism and causal powers. Powers of an entity exist independent of its context and who is studying it.

The identification of a natural kind from other kinds becomes important for explanation if only natural kinds are seen as really existing. Establishing some criteria or practice for demarcating natural and human kinds becomes a vital task. Putnam (1973) and Kripe (1980) claim that empirical research provides the basis for naming and assigning natural kinds. Li (1993) suggests that this merely moves the question of kinds from the realms of philosophical debate to scientific debate without really tackling the basis of the question. Schwartz (1980) similarly holds that natural kinds are defined by traits that are discoverable empirically and tested for by his counter-example test. Scientists identify which traits are associated with natural kinds through study of individual entities. These entities are, however, studied in a manner consistent with their being the member of that kind. This somewhat circular argument makes it difficult to break out of an empirical bind.

Unfortunately, for physical geography the status of entities such as rocks, rivers, and even the events that trigger landslides, as natural kinds, is debatable. Aristotle, for example, only gave the status of natural kinds to a limited range of entities (Loux, 1998), the elements of his day. Contemporary fundamental quantum physics might claim the same status for its entities, but the changing nature of what has been seen as fundamental or elementary entities over the last 2,000 years seems to count against a view of a once-and-for-all answer to the delineation of a fundamental 'natural' kind. Although the existence of geographic entities is not denied, except in extreme interpretations (e.g. van Inwagen, 1990), their explanatory status is often viewed as inferior.

Dupre (1993) and Shain (1993) suggest that the definition of natural kinds is dependent upon the context of enquiry. Li (1993) argues that in naming a natural kind there is always a vagueness at what, precisely, the process of naming is directed. It is only possible to know natural kinds through instances, individual entities that exemplify the kinds. There is no direct access to a natural kind. Naming is based upon experience, both individual and collective, and not upon some absolute knowledge of reality. Identification and naming is based upon a comparison of entities. As the process of comparison can be virtually endless, it is for all practical purposes never complete or capable of completion. This means that the absolute assignment of an entity to a kind can never be certain: I have argued that, because we have no direct access to a natural kind as we do to an individual object, when we name a natural kind exactly what the kind is cannot be determined without further focus. The process of further focusing can never come to an end.

Nominalists, according to Hacking (1983), do not deny the reality of an external world containing entities which interact. They believe, however, that we impose our classification, our own divisions, upon this reality. Hacking (1991) suggests that kinds become defined and important when their properties are important to the individuals who want to know what entities and, by extension, kinds do and what can be done to them. Practice and experience become central and important factors in the process of identifying kinds: When we recognize things to use, modify or guard against, we say they are of certain kinds. Singular properties are not enough. Realizing that a thing has some properties or stands in certain relations prompts belief that it is of a certain kind, i.e. has other properties or stands in other relations. Kinds are important to agents and artisans who want to use things and do things. Were not our world amenable to classification into kinds we cognize, we should not have been able to develop any crafts. The animals, perhaps, inhabit a world of properties. We dwell in a universe of kinds... Natural kinds, in short, seem important for homo faber.

Hacking (1999) identifies two types of kinds that may prove a useful typology in geography. Indifferent kinds have no reaction to being named. A limestone rock does not respond to being classified as a limestone rock. Its nature or essence does not alter, nor does it begin suddenly to behave as a limestone rock when it did not before. This is not to say that we, as investigators, do not interact with the rock in a particular manner because we have defined it as a limestone rock. A stonemason might work the limestone in a particular way and expect a particular response. Scientists may test for the presence of limestone by watching for bubbling from the rock surface after the application of hydrochloric acid. Neither of these behaviours provokes a response from the rock independent of our interpretation and action upon it. Interactive kinds, on the other hand, react to the act of classification. An individual defined as a refugee becomes embroiled in a set of relationships that operate because of that act of classification. The individual reacts to that classification and responds in a manner permitted by or in keeping with that classification.

The act of classification provokes a response. Most, if not all, kinds in physical geography could be thought of as indifferent kinds. There may be an implicit assumption that indifferent kinds are somewhat more 'real', more stable than interactive kinds and therefore a 'better' basis for explanation. Even indifferent kinds, however, are both social and real (Hacking, 1999). The basis for identifying and researching kinds is the dialogue between reality and the researcher. Extraction of properties is always made for some purpose. The distinction is not between some real, indifferent natural kind and some interactive socially constructed kind. There is no distinction to worry about. Trying to establish a distinction is irrelevant. Goodman (1978) identifies relevant kinds as those defined as relevant for some purpose by the user. Goodman emphasizes that relevant kinds assume an external reality with real entities. This reality is, however, only accessible through our actions, guided by our intents. We divide reality up and make entities for our own relevant ends. There is no indication, and no way of knowing, if our divisions of relevance match 'real' divisions.

Kinds in physical geography are always relevant kinds derived from a time- and place-specific dialogue. Their contextual nature makes them no less real in the framework of study of which they are a part. Semiotics, as put forward by Baker (1999), has a similarly blurry image of entities and recognizes their indeterminate status. Entities have to be viewed within the web of signs of which they are a part, as noted in Chapter 2. Entities, and by extension kinds, become contextually defined rather than objects defined independently of study context: As a thing it merely exists, a node of sustenance for a network of physical relations and actions. As an object it also exists for someone as an element of experience, differentiating a perceptual field in definite ways related to its being as a thing among other elements of the environment. But as a sign it stands not only for itself within experience and the environment but also for something else as well, something beside itself.

It not only exists (thing), it not only stands to someone (object), it also stands to someone for something else (sign). And this 'something else' may or may not be real in the physical sense;... Divisions of things as things and divisions of objects as objects are not the same and vary independently, the former being determined directly by physical action alone, the latter being mediated indirectly by semeiosy [sic], the action of signs . . . Divisions of objects as objects and divisions of things as things may happen to coincide... But even when they coincide the two orders remain irreducible in what is proper to them. This quotation provides a useful perspective on the semiotic or pragmatic realist interpretation of entities. An entity takes on a meaning and is only constituted because of its relations in a triad of sign, interpretant and interpreter. The entity exists as a thing,

Species as Natural Kinds

The status of species as a natural entity (Ruse, 1987) for biological classification has been questioned by a number of researchers. Dupre (2001) suggests that the species do not effectively function as a unit for both evolution and classification in general. He makes the point that a unit of evolution needs to be an individual, whilst a unit of classification by definition has to be of a kind. It is the individual organism that is in competition with others of its kind, as well as eking out a living from the environment. According to Dawkins (1978), the unit of evolution is the gene that provides the basic building block for evolution to occur. Organisms are merely differentiated shells surrounding genes. The interdependence of organism and gene is just one problem with this restrictive view of the basic unit of evolution. According to Dupre (2001), evolution occurs when a set of properties that characterize individuals in a phylogenetic lineage change over time.

The lineage is the unit of evolution, not some entity at a single point in time. If this is the case then what properties alter and how does their alteration not change the essence of that species or gene? Both for species and genes the temporal dimension raises problems for identification and classification. The idea that a species can be the same thing throughout its evolution implies that the same species, or rather aspects of it, exists at all stages of its evolution. Such a view implies that each current species is somehow inherent in its past evolutionary lineage. Levins and Lewontin (1985) and Lewontin (1995) highlight that the organism may be the unit of evolution, but it is not necessarily a passive unit. A classic view of evolution implies that the environment sets problems and organisms find solutions. The environment is fixed; organisms are malleable. The organism is a passive respondent to changes in the environment. The advent of genetics may seem to imply a more active role for the organism, but instead it has emphasized the internal, inherent characteristics of the genes. The organism is still a passive unit; the genes alter so the organism has to. The organism becomes almost an infinitely deformable conduit for the external influences of the physical environment and the internal influences of its genetic make-up. Levins and Lewontin (1985) and Lewontin (1995) instead suggest that the organism is an active motor of change and, at least partially, a determinant of its own destiny. They base their claims around a number of key points.

First, organisms determine what are relevant characteristics of the environment for them. The concept of a simple niche model to reality is difficult to sustain. Adaptation to fit a niche requires defining the niche before the organism exists. In a similar vein, organisms can create their own microenvironments. As Levins and Lewontin (1985) note, even without manipulation of their surroundings, all terrestrial organisms have a boundary layer of warm air resulting from the metabolism of the organism. Second, organisms alter the external world that they interact with. Plants alter the substrata they are on by their activities, and alter the soil structure and chemistry locally. Third, organisms interpret the physical signals they receive from the physical environment. Any environmental change is mediated to the organism through its senses, and its response will depend on what it senses and how it interprets this change. Fourth, organisms can alter the statistical pattern of variation in the physical environment. Changes in air temperature may be dramatic, but an organism with feathers does not sense anything other than extreme changes in temperature.

Lastly, it is not the environment that defines the traits to be selected by evolutionary processes, it is the relationship between the organism and the environment that defines these. The organism is an active participant in the environment, and it is how the organism relates to the physical environment, other organisms and its own species that determines which traits are of use and which are not. There is no predefined set of traits that are inherently better than some other set of traits; it all depends on context. The above points illustrate that the unit of evolution is difficult to isolate from its relations. Likewise, the organism as a member of a kind needs to be thought of as an active rather than passive agent. Its properties are context-dependent and subject to change, but these properties are present, and therefore have the potential to change, because the organism is a member of a particular kind. Activation of these properties is not inevitable, it all depends on the relations an organism has in a specific context.

The above discussion implies that classification of species is fraught with problems. Dupre (2001) suggests that classification, when associated with a central theory, can produce a 'real' classification of reality, even if it is, he believes, a weaker claim that is traditionally assumed for natural kinds. Kinds themselves, however, are not the sort of things that exist in reality. Instead it is the members of the kind that undergo the effort of existing in reality.

MAGNITUDE AND FREQUENCY – ENTITIES OUT OF CONTEXT

Wolman and Miller (1960) outlined the significance of magnitude and frequency of events in geomorphology by abstracting events from their context of operation. Measuring an input, an event, to the system and then its immediate output or effect, they could construct a relationship between the two that was meant to hold for any system anywhere.

Events were defined as discharges of given magnitudes, whilst output was defined as the sediment load carried by the river. The theory they developed, or 'truism' as Haines-Young and Petch (1986) describe it, was that middle-size events, normal events, did the most work in a catchment. Small events were frequent, but of insufficient power to move material. Large events moved a great deal of material, but were not frequent enough to move material with sufficient regularity to sculpt the landscape. It was the middle-size events (bankfull discharge events in their study) that occurred with sufficient frequency and force to move material around and out of the catchment so as to sculpt the landscape.

Lane et al. (1999) noted that this initial description of the relationship between magnitude and frequency relied heavily upon the idea of a 'normal' operating mode for the fluvial system within a catchment. The idea that the fluvial system tended towards an equilibrium form with a 'normal' mode of operation was an important basis for legitimating the abstraction of events and effects from their context. Abstraction was permitted, because the events and effects were illustrations of this normal mode of operation. Event characteristics were reflections of an ideal event type that had a specific and contextless relationship to an equally ideal effect. This meant that the relationship between event and effect could be modelled as a simple input–output relationship that could be transferred between fluvial systems. The magnitude–frequency curve derived could be applied to any system regardless of its environmental context. Subsequent development of the magnitude–frequency concept by Wolman and Gerson (1978) gave some recognition to the context dependence of the abstracted relationship. They introduced the idea of effectiveness to describe how different landscapes may respond differently to the same event, the same input.

Effectiveness related the magnitude of the change caused by an event to the ability of the landscape to restore itself. Both parts of this equation were context-dependent. Within a temperate environment an event of a given magnitude will have an impact of a specific magnitude. The temperate environment will tend to recover from this impact and move back towards its 'normal' state. In a semi-arid environment, the impact from an event of the same magnitude will be greater and the time required for recovery will be longer, possibly even longer than the time intervals between events producing the same impact. Although context is now introduced into the analysis, it is still a very limited view of context. Likewise, the system still retains a movement towards equilibrium or 'normality' as its functional goal.

Lane and Richards (1997) highlight the need to distinguish between the 'immanent' ahistorical processes and the 'configurational' (Simpson, 1963). The former are general processes that always occur under specific conditions, whilst the configurational represent the result of the historical interaction of the immanent with particular historical circumstances. This mirrors the complicated relationship between event magnitude–frequency and impacts. The relationship can be abstracted and general properties derived. The relationship is a generalization, often a statistical caricature of complicated and temporally distinct relationships. The relationship cannot be understood as an ahistorical process. Generalities are only derived from particular circumstances. Without knowledge of these circumstances the generalities cannot be related back to reality. General properties are only discernible because the processes that cause them operate within specific contexts derived by unique historical sequences of change. an object and sign simultaneously, and each of these elements is interdependent and uninterpretable without the others.

The sign, the focus of study in semiotics, is the 'something else' of the quote. The sign is constructed through interpretation by an interpreter of an object. If the object and thing are identical the sign is an interpretation of reality as it is. The coincidence can never be known for certain, but it would also be inappropriate to assume its presence meant an object could be reduced to a thing or even vice versa. The nature of a thing is associated with its relations to a physical reality, relations beyond the knowable for a researcher. The thing is wholly or partially related to an object that is constituted by relations to the researcher. This latter set of relations means that the object is constituted via signs; it stands as an outcome of interpretation. This means that its very nature is different from that of a thing, the reality, that it is a representation of. Any knowledge a researcher can glean about reality is via the object and so only knowable via signs, via interpretation. Knowledge without relations to signs is not possible.

Entities are not reducible to some common baseline of reality, they are only reducible to objects of study, with all the difficulties of being part of a dialogue that this implies. In some ways this interpretation of the individual entities is similar to the relational view of entities put forward by Whitehead (1929). He understood entities to be capable of distinct definition, but only by virtue of an entity's relations to other entities and, in turn, their relationships to the wider whole of reality. Any single entity at a specific time contains within it, its relation to itself in the past and future. Past relations express the constraints on an entity, whilst future relations, as yet unrealized, express the potentialities of an entity. In Whitehead's view all the relations that come to define an entity in the present are the constraints of its past and explain why it is as it is. The uncharted future of an entity is not random nor completely expansive. An entity's future is constrained, but not singularly determined by its past (what it is).

The nature of an entity defines its possible futures, its potential relations. In this way, an entity embodies both what it is, what it has become and what it could be or is becoming. There is, however, no need for an entity's nature to be rigidly fixed or stable over time. Entities can vary in their spatial and temporal stability. Whitehead's vision of entities as observer-defined abstractions, solidifications, of the flow of processes, captures the distinct, but ever-changing, nature of an entity. Entities require flows of processes, their relations with other entities and the wider whole of reality to continually reproduce themselves. Stability of these relations provides the appearance of solidity, of illusory permanence. Semiotics, with the continual creation and destruction of networks of signs that constitute an entity, is a view in a similar vein.

The semiotic approach means that both entities and kinds do not need to have any physical reality, although they will have relations to this reality. In semiotics and pragmatic realism there are no natural kinds or entities – neither exists independently of an interpreter: And the metaphysical categories that emerge through this endeavour yield an understanding of nature or the dynamical object which, as an indefinitely rich evolving continuum that must be made determinate for our awareness by the manner in which we 'cut' into it, cannot provide the basis for final ultimate knowledge. Nature or the dynamical object, with its qualitative richness, lawful modes of behaviour, and emerging activities, constrains our interpretations, pulling them or coaxing them in some directions rather than others. It answers our questions and determines the workability of our meaning structures, but what answers it gives are partially dependent on what questions we ask, and what meaning structures work are partially dependent upon the structures we bring.

Thus, within this interactive context of interpretation and constraint, different structurings yield different isolatable dynamical objects, different things, different facts. Truth is always related to a context of interpretation. This is not because truth is relative but because without a context of interpretation the concept of truth is senseless, indeed literally so. Knowledge involves convergence, but convergence within a common world that we have partially made, and continually remake in various ways.

As the quotation points out, the divisions of reality, the kinds studied, are of our making. We do not, however, start with a blank page. All our interpretations are constrained by an underlying and independent reality. We interrogate that reality and kinds emerge as an outcome of that interrogation, but are not existent independent of that interrogation. Entities of study, likewise, are dependent upon this interrogative process. It is of vital importance, however, to realize that the dependence of kinds, entities and objects upon interpretation does not mean that they reflect a purely relativistic version of reality. Kinds and entities are not relativistic in the sense that there are an infinity of other possible interpretations, all equally valid, as there is no basis for assessing the validity of each. The basis for assessing the validity of entities and kinds constituted is the interpretative context of their constitution. It is only relative to this context that appropriate methods of validation can be constructed. It is the interpretative context within which judgements concerning appropriateness can be passed.

The interpretative context, like the entities and kinds within it, is open to continual renegotiation, to continual alteration and evolution. Many scientists would claim that this renegotiation is the convergence mentioned in the quotation towards a 'truer', in the absolute sense, version or interpretation of reality. This assumes a progressive direction to the alteration of the interpretative context of which we can never be certain.

The continuation of certain entities and kinds with changing interpretative contexts testifies to their adaptability rather than proof of their ultimate reality. A more appropriate metaphor for alteration of interpretative contexts may be that of evolution. Changes in interpretation in the interrogation are appropriate for the needs of the researcher in that time and place. These needs, and the alterations they cause, constrain the nature of further alterations in interpretation; but how is unclear, because, in the present, we work only with the singular, historically unique and constrained outcome of the past.

Classification

The above discussion of entities and kinds provides the basis for dismissing the idea that there are natural classifications in reality. Division of reality is undertaken by researchers working within a unique interpretative context with associated versions of kinds and entities. The above discussion suggests that these kinds and entities do not correspond to reality as it is, but rather to reality as a useful framework for the researcher. Classification practices reflect this view of reality. Classification is based on usefulness to a researcher rather than determining the absolute structure of reality. Classification of reality, therefore, becomes a means to serve the researcher or group of researchers' ends. Classification is a research tool like any other; it is an aid to interpretation rather than an absolute statement about the nature of reality. Decisions about how to divide up reality, which at heart is what classification is, require the use of some property or properties associated with entities and kinds.

There are two ways in which properties can be used to classify reality: by lumping things together and by splitting reality apart. Lumping and splitting are two diametrically opposed ways of looking at reality. 'Lumping' views properties as being a cumulative means of combining entities into a single reality. Properties should be chosen that help in linking diverse entities. Commonalities are searched for. 'Splitting', on the other hand, looks for properties that enable an initially undifferentiated reality to be divided into distinct, unique parts. The two approaches to classification could be seen as bottom-up (lumping) and top-down (splitting), the former working from the level of individuals up, the latter down from the level of the morass of an undifferentiated reality. Choosing whether to lump or split can also influence the seemingly objective choices of quantification of classification. Foley (1999), for example, in relation to analysing hominid evolution, noted that multivariate statistics tend to highlight overlaps between hominid morphologies whereas cladistics operates most effectively when differences between morphologies are the focus of study.

A key question is which properties to select to differentiate reality, whether lumping or splitting. Property selection for classification may be related back to the theoretical framework used. Theory defines what properties reflect the essence of entities and which are contingent. Theory, in other words, guides the researcher and defines which properties to use for classification of the 'real' entity or kind. Ideally, properties that are essential for the entity to exist would form the basis for classification. These properties are in combination associated only with that entity and so provide a clear and unambiguous means of separating the entity from the rest of reality.

Such property identification and association is rare. Often classification requires not properties unique to an entity but properties that vary between entities. These can be measured over a range of values and entities assigned to a class based on their position in this limited continuum. Assignment requires division of the range of properties into specific zones. If a property value falls within that zone the entity is assigned to that class, if the value is outside the zone it is assigned to another.

Many properties can be used in combination to provide a more complex, composite measurement scale, or rather multi-dimensional measurement space in which to classify entities. The principles are, however, the same as for a single index. The measurement space is divided into discrete volumes. A composite value within the range of a specific volume means assignment to that class. The issue is more complex than just identifying properties of essence, however. Often properties useful for the classification cannot be measured. They may be properties that exist in theory, but which have no clear or tenable translation into measurable properties. In these cases, surrogate properties that are measurable and assumed to be related in some manner to the unmeasurable theoretical properties are used instead. The problem is that there is not usually a one-to-one relationship between measurable and theoretical properties. A measurable property may only reflect a particular aspect of a theoretical property. Similarly, a property may be illustrative of a number of theoretically derived properties. A simple measurable property may reflect a whole host of theoretical properties to varying degrees.

A lack of a one-to-one correspondence does not provide an ideal basis for classifying entities in an absolute manner. The implication is that all classification is conditional, although some classification schemes seem to be more stable than others. The Linnaean division of the living realm, as mentioned earlier, seems to be a successful and stable classification scheme. Stability does not, however, equate with reality. Just because a classification has been stable or appeared stable does not mean that we can say for certain that its divisions divide reality as it is. As with entities and kinds, classes are convenient and relevant divisions for researchers. Classification using measured properties requires another factor: standardization. Classification, particularly in the natural sciences, requires the repeated measurement and consequent consistent assignment of entities to a specific class. Classification would expect or demand that an entity be assigned to the same class by independent observers. Property measurement and class allocation need to be consistent across space and time to permit such comparison and assignment. Without confidence in the classification procedure providing consistent assignment, researchers could not be sure that they are studying and discussing the same entities and kinds. Without this confidence interpretation of phenomena would not be a focus of analysis; rather, it would be the validity of the entities used.

In this context the process of standardization of property identification, measurement and use for assignment ensures that definitional quality is maintained or guaranteed. Standardization, however, requires the implementation and co-ordination (convergence and alignment in actor network terminology (ANT)) of a complex social and material network. Indeed vast volumes of papers have been written to justify a particular classification system and the particular methods of measurement required to implement it. The example of soil classification provided below illustrates how classification schemes require detailed and often committee-based agreements concerning the precise methods to be used and the interpretation of results. Such concerns are not minor details amongst important theoretical or philosophical discussions.

Given the central importance of standardization to the investigation of physical phenomena, the life blood of physical geography, it is a surprise that it is not considered in more detail in the appropriate literature. Procedures of standardization are often so entrenched within the traditional practices of subject that they pass without comment or are relegated to methodological footnotes or dismissed as irrelevant methodological niceties. How agreed and common procedures for measurement arose, and the assumptions and often highly personalized battles that classes embody, are lost in their routine application. The discussions, the compromises, the accepted exceptions and the basis for redefinition of the standard are not part of the usual publication diet of physical geography. The sociological, and therefore, by implication, subjective and relativistic connotations are not seen as part of an objective study of physical phenomena.

Classification of Soils

Soil classification is essential to help clarify a complex phenomenon. Although most authors accept that classification is imperfect some believe that certain types of soil classification are superior to others: If, however, the criteria of grouping are based on intrinsic properties not specifically linked to the objective [of the study or activity] it is hoped to achieve they may be called 'natural'. For the broadest use of 'natural' systems of classification are likely to be more generally helpful and therefore should be based on fundamental properties of the objects in a natural state. The quote illustrates that Townsend views intrinsic properties as the key determinant of a natural versus artificial classification. This is a common view within soil classification. Inherent properties are viewed as objective criteria, yet defining what is inherent is difficult. Properties derived from the parent material within which soils form could be viewed as inherent, but these alter as the soil develops.

Assuming soils progress through predefined stages of change could provide a basis for defining inherent properties based on the development stage. This argument could become circular, however, as the stage will be defined by the properties. Another common belief illustrated by the quote is that a 'natural' classification will be superior to any other and by implication form a sound, objective basis for the definition of other, more purpose-orientated classification schemes. The implication is that somehow the 'natural' scheme proposed as the ideal has no purpose! Other authors, however, take a more pragmatic view of soil classification: [A]ny classification must have a purpose and with soils the aims can vary widely. For the most part, general systems developed for the definition of map units in soil surveys designed to aid land use will be discussed, but others have been constructed with less practical aims and more emphasis on soil formation and evolution to summarize relationships for scientific purposes.

Classification is justified as long as it is related to a specific purpose and is consistent and coherent with that purpose. Whether this same scheme will be of use for other purposes is open to debate even if the properties being used for classification are similar. This has important implications when classification schemes are used to produce products such as soil maps for agricultural productive purposes. If the classification schemes, no matter what their purposes, are referring to the same phenomena in reality, then overlaps or commonalities between schemes may be expected. These arise not only because the theory underlying identification and description of the phenomena inform both schemes, but also because theory also informs the properties to be measured and their means of measurement. This means there may be convergence in the characteristics considered in classification schemes despite their different purposes.

Clayden (1982) makes the point that there are two potential ways of classifying soils. Hierarchical classification can be used. This framework has a 'family tree' structure with an initial 'type' defining membership of the class in terms of a range of properties and finer divisions being made within these properties to refine the classification. An example of this is the USA soil taxonomy developed by the Soil Conservation Service of the US Department of Agriculture. Soils in the hierarchical system are divided on the basis of precisely defined diagnostic horizons and soil moisture and temperature regimes. Soils are then classified into orders, suborders, great group, subgroup, family and series.

An alternative means of classification is the co-ordinate classification system. In this system entities are still differentiated, but the properties used are not necessarily arranged in a hierarchy or even ranked. The former classification type is, according to Clayden, easier to remember as properties are prioritized in importance. Although the latter classification scheme has been used in Russia, East Germany and Belgium, the number of classes that are needed and their unstructured nature has meant that these systems have not been taken up internationally. Classification systems such as employed by ISRIC (1994) are the result of a long process of negotiation between usually national agencies. The first FAO/UNESCO soil map of the world began in response to a recommendation in 1960, with the first maps being published in 1969. From that date numerous meetings have been held to develop a common basis for classification, a common nomenclature and common methods of analysis.

The result is not an agreed and single coherent framework based on a unifying theory or agreed hierarchy. Instead the product is a compromise between existing systems developed for national and even regional interests and a need for international clarity in what researchers are referring to. The result is a classification scheme full of familiar local names such as chernozems and kastanozems, mixed with rigorous methods of demarcation. Clarity is provided by the use of two sets of properties for defining soils: diagnostic and phases. Diagnostic horizons are used for identifying soil units, and their presence is based upon the operation of specific diagnostic processes in the soil. Diagnostic properties are soil characteristics that do not produce distinct horizons, but which are of importance in classification. Terms such as 'andic properties' (exchange capacity dominated by amorphous material) and 'hydromorphic properties' (water saturation conditions) reflect such properties. Surface diagnostic horizons are defined so that their main characteristic properties are not altered by short-term cultivation. This is an attempt to permit the application of diagnostic horizons to lightly cultivated areas. Phases are features of the land that are significant for its use and management. These features enable land use and its impact to be considered in the classification.

Significantly, it is the soil in its 'natural' uncultivated state that is seen as the basis for classification. Soil altered by cultivation is viewed as a secondary or inferior entity with secondary properties. This reflects a general attitude in soil classification schemes, 'natural' first then modifications to the natural. Natural relates back to the various theories embodied by the different classification schemes the FAO/ UNESCO classification tries to accommodate. These theories reflect a concern with understanding the causes of soil formation – primarily the role of climate. Latitudinal and altitudinal variations in climate are matched by variation in soil properties and modes of development. Deviations from these 'model' forms are viewed as oddities within specific developmental sequences, and are treated as such.

SUMMARY

Entities are the centre of any study of reality. Entities are the units that we believe exist and which we can manipulate and use for testing ideas. Entities have properties or attributes associated with them, but not all of these are essential for defining the essence of an entity. Entities and kinds are intimately related. Kinds provide a template for what it is to be a member of that kind. They dictate the essential properties that define a member of that kind, as well as the behaviour to be expected. The kind becomes the focus for explanation; the entity becomes an illustration of the kind. There are different types of kinds and it has been argued that research should focus on natural kinds. Even if geographic natural kinds exist, all kinds are defined for a purpose. This means that all kinds are to a large extent observer-defined and defined for a purpose, not as some natural building block that forms the 'real' basis for explanation. Kinds and entities are enmeshed within a network of signs and their meanings are likely to evolve as their purpose alters. Hacking (1999) identifies two types of kinds: indifferent and interactive. Physical geography deals with indifferent kinds; that is, those that do not respond to the act of classification.

4

FORMS OF EXPLANATION

EXPLANATION IN PHYSICAL GEOGRAPHY

This chapter outlines and justifies the different types of explanation that are regarded as acceptable within physical geography. Central to each mode of explanation are the researchers themselves. By defining the nature of the study, the questions to be asked and the entities to be studied, the researcher also sets the criteria by which a study is judged to be successful. Despite the wide variations in underlying philosophies, what is regarded as an acceptable explanatory framework needs to be established in order to understand where certain philosophies may be more appropriate for certain types of explanation and subject matter.

What is Explanation?

This question is a difficult one to answer despite its seeming simplicity. Although it is difficult to separate explanation from its philosophical framework, a general discussion of what explanation is may help to provide an outline of what physical geographers would find an acceptable explanation. Explanation can be summarized as providing an answer to the question 'why?' – or, more specifically, why is something as it is? Given what we know about why, what should we expect the something to do in the future? In answering these questions our explanations reduce uncertainty about reality and the future of our reality. Explanation and prediction are not the same, but are related to each other. Explanation could be viewed as a need to make the unexpected expected, to reduce the amount of surprise that anyone encounters. Surprise is replaced by expectation. When expectations are confounded by the unexpected, people tend to want to know why (Toulmin, 1960).

The impetus for explanation may be that simple. Some arguments could be constructed about the evolutionary advantage such a search might provide, but such speculation is not really a concern here. An argument could also be made that replacing the unexpected by the expected reduces individual and social stress. By reducing unexpectedness, explanation could also be viewed as beginning to represent reality as it really is. Stating that something has an explanation is tantamount to stating that we have been able to identify what reality is and from that identification been able to understand how reality works. Explanation and understanding of reality come together. Explanation implies that a mirror has been held up to reality and an undistorted, or at least acceptably distorted, reflection obtained. Whatever the case, the impetus immediately pushes the interaction between an individual and their surroundings to the fore. The need for explanation is based upon interaction; it is based on a need, an impulse to know about the environment and its behaviour.

The impetus for explanation need not be the same for different individuals. The example Toulmin uses is that of a stick appearing to bend in water (see Harvey, 1969). For some people this is a 'so what' moment. The significance of the interaction with reality is of no concern to them. For others, it poses questions about their reality. For the latter group of people, explanation provides an insight into how reality really is. Explanation produces a possibility of organizing reality into simple categories – what is explained and what is not. Additionally, these categories can be refined into, for example, what explanation A explains and what it does not, what explanation B explains and what it does not, and so on. In other words, a systematic body of explanations can be built up and used to define which parts of reality, or rather our interaction with reality, are explained and which are not. Viewed in this manner explanation needs to be contextualized. Without some way of classifying reality, without a particular means of interacting with reality, the sources of unexpectedness that provide the basis for explanation are unknown.

Training and experience in what to expect and what is unexpected, and the significance of each within one's frame of reference, needs to be identified. In this manner explanation, theory and philosophy are interwoven. Accepting an explanation as valid implies that there is a set of criteria by which an explanation is judged. Once again the context of explanation becomes important. What is thought to be a valid explanation tends to depend upon the rules and conventions of a particular group. This could be interpreted as saying that explanation is socially determined. The exact rules may depend on social factors, but it could be argued that the need to predict expectations of reality from explanations means that the rules must converge to provide a version of reality adequate for the purposes of that group. This could be taken to mean that rules for an acceptable explanation should be rules that permit explanation to reflect reality as it is. This would severely limit the nature of the rules that could be applied.

Hempel (1965) suggested that scientific explanation was concerned with both universal and statistical laws. Universal laws are general principles, possibly derived from observation of regularities in reality. Given an initial set of premises, universal laws are applied to these to deduce an outcome logically. This is the classic covering law model. Universal laws usually imply that we have an idea of the reasons why premises and outcomes are linked, that we have some mechanism in mind. This mechanism will always produce the outcome, given the premise. Statistical laws can be applied in the same manner to premises and outcomes; the covering law model can be used, but we may not be certain that we will always get the outcome. The laws may be statistical because we do not understand the mechanisms involved or because they reflect observations of common conjunctions of premises and outcomes. Statistical laws, unlike universal laws, are only true because we observe them with our current level of knowledge; they have the potential to change, unlike universal laws.

Causality

An important feature of any explanation is the idea of causality. At its heart the principle of causality assumes that every action, every event observed, has a cause. The event is determined by the cause. An event is as it is because it has been determined by events before it in a long chain leading back to what a researcher would identify as the cause. This means that physical geography and physical geographers tend to believe that events they observe are determined, that there is some knowable explanation, some knowable cause (or causes).

The simplest view of causality is the idea of cause and effect. An event happens prior to and probably close to another event. The first event can be interpreted as the cause of the second event – the effect. Cause and effect imply temporal and spatial proximity, although their linking may also require some ideas of why they are linked – a theory.

Simplistically, there are two views of causality: the successionist and the generative. In the successionist a cause is merely something, an event, that happens before something else. Connecting one event to another is something humans do, but the connections are merely our constructions and have no existence in reality. In this view of cause, the entities that form the events that occur in succession are passive objects. They do nothing to influence each other; they do not interact in any manner to produce events. The generative version of causality takes the opposite view. In this view an event is generated by a cause and so the two events are linked and dependent upon each other. Taken together, cause and effect could be viewed as forming a single event, they cannot be considered separately as single events. A specific cause must generate a specific effect of necessity (Harre, 1985).

In this view it is the nature of the entities themselves that generate the relationship. The nature of the entities involved in an event determine that event. An entity, by its very nature, being of a particular type, must possess particular properties that determine the manner in which it must, of necessity, behave under a given set of circumstances. Abstracting causality from the entities involved can produce information about general relationships, but it also tends to highlight the successional view of causality as cause is displaced from the actual entities involved in generating the succession.

The choice between these two views of causality has been the core of much metaphysical debate (e.g. Cooke and Campbell, 1979; Holland, 1986; Oldroyd, 1986), but the debate largely boils down to issues considered in previous chapters on the nature of reality and human ability to understand reality as it is. Hume viewed cause and effect as a succession of events linked together by humans by an assumed mechanism. This was a human construction based on experience, not a real structure of reality. Rescher (1991) views causality as part of our conceptual framework for bringing shape and order to a formless reality. The researcher is the centre of the construction of causality. A generative view of causality sees events as linked of necessity and therefore reflecting reality as it is. Entities possess the potential, depending upon circumstances, for causing effects. Causality is about real relationships between entities and their actions. The key problem is that humans have no absolute knowledge of reality, no privileged position from which to view things as they really are.

Kant tried to resolve the two views by invoking two realms of existence, the noumenal and phenomenal worlds. The noumenal world is reality as it really is. Humans do not have direct access to this world, instead all they can know is the phenomenal world, the world of phenomenon as constructed by humans. Causality then becomes a messy mix of the two. Causes and effects are generated by entities and their relations, but all we as humans can know are the succession of events. It is this succession of events upon which we impose an interpretation based on what we believe to be the 'real' relations between 'real' entities based on 'real' laws. All the 'reals' being human constructed, but constructs based upon a participation in reality, an active structuring of phenõmenon. Bunge (1962) identified three meanings of causality. First, causation which associates a particular event with a particular result (cause–effect). Second, the causal principle which casts law in terms of cause and effect. Lastly, causal determinism; that is, a doctrine that asserts the universal validity of the causal principle.

The three meanings are not unconnected. Assuming that everything is explainable by cause-and-effect relationships implies acceptance of a causal deterministic view of reality. It implies that cause and effect are observable because of the underlying importance of laws in determining how reality works. Within science causality has a more restricted scope and definition. Humean cause is based on constant conjunction (Peters, 1991). Cause is always accompanied by an effect, with the cause occurring before the effect. This conjunction was believed to be sufficient to provide a basis for establishing a causal link. Such constancy in conjunction is difficult, if not impossible, to achieve in a field science, so more rigorous concepts are required. The operation of laws ensures that reality is composed of events that are causally linked both spatially and temporally: cause-andeffect relationships.

The presence of cause-and-effect relationships can be interpreted in terms of necessary and sufficient conditions. Any event could be thought of as having a set of causes associated with it. These causes are the conditions required for an event to take place. A potentially infinite set of conditions is not capable of study, so a subset or subsets of conditions are the focus of study. These are the necessary conditions and the sufficient conditions. Necessary conditions refer to the conditions that would justify the nonoccurrence of an event, whilst sufficient conditions are the conditions that would justify the occurrence of an event. Necessary conditions are more restrictive than sufficient conditions. Necessary conditions define the set of requirements that must be present if a certain event is to happen. These conditions can only be identified, however, if the event is prevented from happening by these conditions not being present. They imply an ability to manipulate reality to prevent the necessary conditions to allow their identification, a luxury that is not usually possible in field sciences.

Cause-and-effect relationships reveal the laws that govern reality. Underlying any cause-and-effect relationship is a set of structures. The operation of these structures ensures the consistency of causes and effects being observed. Where cause–effect relationships are not observed these underlying structures either are not present or their operation is influenced by other structures. Identification of the underlying structures can be useful in identifying expected causal relationships.

NECESSARY AND SUFFICIENT CONDITIONS

Simms (2002) identified a series of karst forms around the shores of Irish lakes that had a very specific distribution. The rohrenkarren or tube karren are vertical upward-tapering, closed tubes. In a detailed study of the distribution of these forms Simms found that the three lakes he used in his study, Lough Càrra, Lough Corrib and Lough Mask, all shared a common feature: all were in a near-permanent state of carbonate saturation. At all three active precipitation of carbonate, as exhibited on the encrustation of plant stems and mollusc shells, was observed. Lakes that were significantly undersaturated did not show any tube karren formation. This means that carbonate saturation is a necessary condition for tube karren formation. The absence of this state means that this karst form will not be produced whatever the other favourable conditions for its formation. Identification of this necessary condition relied upon knowledge of the saturated state of other similar lakes, such as the Killarney lakes of County Kerry. Saturation was not itself, however, a sufficient condition for the formation of tube karren. There needs to be a set of sufficient conditions as well, conditions that produce the form.

Simms used his observations of the distribution of tube karren to advance a possible mechanism for their formation. Tube karren were confined to the epiphreatic zone produced by seasonal lake fluctuations. The level of fluctuation varied between lakes from 1 to 3 metres in tandem with the magnitude of tube karren found. Tube karren were formed by the condensation of water vapour in the air-filled apex of the tube. This water vapour is not saturated, unlike the lake water from which it is derived. Condensed water dissolves limestone at the apex of the tube, but dissolves less as its aggressiveness Water condensesat apexSolution decreases in aggressivenessas it runs down side of tube is reduced by reaction as it runs down the tube. This means that the base of a tube expands at a lesser rate than the apex, as limestone at the base is dissolved at a slower rate. The almost perfect circular cross-section of the tubes reflects the uniform condensation of water onto the side of the tubes and the uniform dissolution that results.

When the condensed water reaches the lake water there will be enhanced precipitation of carbonate. Carbon dioxide will be released by this precipitation that enriches the atmosphere in the tube apex. More carbon dioxide is able to dissolve into the condensing water film and so increases the aggressiveness of the water film. Associated with these sufficient conditions is a necessary condition, the absence of which would mean that condensation and subsequent dissolution could not occur. For condensation to occur there needs to be a temperature difference between the rock and the water vapour. If both the rock and the water vapour were at the same temperature then condensation would not occur. The temperature of the limestone fluctuates at a higher frequency and amplitude than the lake water. This results in the two systems being in disequilibrium, at least for a few hours, when the rock is at a lower temperature than the water.

Simms also suggests that the initiation of tube karren may require a set of necessary conditions. Calm water is essential for trapped bubbles to initiate the indentations that can develop into tube karren. For such indentations to survive the lake water must be at the point of carbonate saturation. Only under these conditions can the minor irregularities persist without dissolution to form tubes. The Killarney lakes of County Kerry, for example, had sufficient seasonal fluctuations of lake levels to permit the above mechanism to operate, but all were undersaturated so the initiation of tube karren could not occur.

Causal relationships can be represented by linkages between causes and effects. At the simplest level is the single cause-and-effect relationship. In this relationship cause precedes effect and the link is assumed to be the causal mechanism. The link itself, however, could be viewed as being composed of a series of smaller events that link the cause to the effect. Linking a rainfall event to a landslide occurring may seem like a simple cause and effect. The link between a rainfall event and a landslide could, however, be viewed as being mediated through smaller and shorter duration events, such as a rise in porewater pressure. When this series of links is observed over a number of particular instances of the phenomenon called a landslide, then the causal structure obtains some sort of stability, at least in the mind of the researcher. A problem is, however, that the causal chain perceived is capable of further and further refinement; it is capable of infinite regression, reducing causation down to the lowest level.

The causal chain may be assumed to be linear and restricted, i.e. a single causal pathway. This is unlikely in reality. It is more likely that a single event at the start of the chain will propagate a number of subsequent events. In other words a single cause need not generate a single event, and nor need the links between a single cause follow a singular path to a particular event. There may be intervening factors that produce multiple potential pathways to an event or even series of events. A rainfall event may generate porewater pressure sufficient to cause a mass of regolith to move, but it may also provide a lubricated layer over which the shear plane can move. In addition, a rainfall event may weather material in the mass of regolith and thus weaken the coherence of the mass, increasing its susceptibility to movement. The single causal event can have a multiplicity of effects, all of which can contribute to the landslide, possibly even at a range of different temporal and spatial scales.

This illustrates that once intermediate factors are viewed as part of an interrelated mesh of effects, the idea of a simple cause–effect relationship disappears. Instead, activation of different possible causal pathways becomes a concern porewater pressure and rainfall may be all that is required to explain the occurrence of a landslide, the other effects are no less real and still contribute to the timing of the landslide. A possible alternative view of cause-and-effect relationships is to view these as a causal web. Causes produce a range of effects that in turn function as a range of causes for other effects. Very soon a complex network of relationships can be built up between events with an associated range of outcomes. Tracing which of these relationships is necessary and which are only sufficient for any particular outcome becomes increasingly difficult as the complexity of the web grows.

Explanatory Frameworks

Given that explanation is the goal of investigation in physical geography then we need to look at how causal relations are put into an explanatory framework. Chapter 2 outlined the basics of deductive argument, and this style of argument is useful for some causal webs. The deductive argument provides a means of linking a cause to an effect through the intermediate stages of effects. Each stage could be seen as a logical outcome of the previous stage and so form a string of logical conclusions and premises that lead from cause to effect. Clearly, the problem of infinite regress of cause has to be addressed at some point; that is, at what level of reduction do you draw the line for an adequate explanation? Despite this problem, deductive reasoning provides a valid and formal basis for explanation. Deductive reasoning provides a basis for identifying what must be the case logically. The logical basis of the links is provided by reference to a 'law'.

The law provides the reason why the cause and effect should be linked. It provides the logical basis for the effect. If C, the cause, happens, then the law derived from an overarching theory predicts that E, the effect, must happen. Put another way, if a state of reality exists at time t0 which is knowable to the researcher, then applying law L, derived from a specific theory, predicts that at time t1 there should be another state of reality. An outcome state can be predicted from the initial state, if a law is applied to that initial state. There is a problem, however, with arguing, or rather presenting, cause and effect in this manner. Using 'state of reality' or 'state of affairs' as shorthand for the situation before and after a law has operated is very unspecific. Cause seems to be located within the phrase 'state of affairs' rather than in an object or entity. The phrase implies that had the state of affairs been different then the outcome may have been different. Questions of how different, and therefore what the role of any individual entity has within this explanatory framework, are not addressed by the phrase 'state of affairs'. Induction, according to von Engelhardt and Zimmerman (1988), refers to the situation where the controlling or initial state of affairs is known, as is the resulting state of affairs. What is unknown is the link between them, the theory and derived law that can connect the two states. Rather than arguing for a link between two states logically, induction infers the link between the states; there is no logical necessity for the inference to be true. All the problems of induction outlined in Chapter 2 are still relevant to this definition. If causation is located within entities and relations, inductive reasoning is arguing that if cause and effect are known then the link can be derived. The manner in which it is derived is unclear, and how the derived link can be assessed is likewise unclear. Any link put forward as a likely candidate can, presumably, be argued for within some context, or else it would not be put forward. If there were no recourse to information about cause and effect other than that contained in the initial statement of positions, then there would be no way in which to compare candidate links. It is only the possibility of other means of assessment, or using other information, that provides a way of assessing candidate links. Once the researcher moves into this form of analysis, however, it could be argued that they are moving from pure inductive reasoning into a fuzzier arena of investigation and explanation.

There is a third style of explanation. Often the effects are known and there is a link, a law that produced them. The unknown is the initial state of affairs, the cause. The argument works back from effect to cause, from resulting state of affairs to initial state of affairs. Unlike induction, there is a possibility of applying logical argument to the analysis. The law can be applied to the effect and a cause predicted in retrospect; in other words, retrodiction can be used as the basis of argument. Given an outcome produced by a given process, we should be able to infer what the initial situation was. This style of argument is called abduction. A key limitation of abduction is that there is no logical necessity for the initial state of affairs to have existed. Unlike deduction, the logical final statement is not a necessary one. Only if the law assumed to link cause and effect, initial and final states of affair, really operated can the argument be logically watertight. The initial state of affairs is the unknown, so there can never be certainty that the law operated – only a likelihood.

There could be other laws that could be applied that suggest that other initial states of affairs could have been the initial state. There needs to be a way to assess which of the possible links back to a range of possible initial states is the most likely. In other words, abduction requires additional information in order to help in selecting what is viewed as an appropriate explanation. From the above discussion it should be clear that only the deductive argument produces a link between cause and effect that arises out of logical necessity. There can be no other effect, given the cause and the law operating. Admittedly, unless the deductive argument is somehow linked to reality, unreal statements can be logically deduced from inappropriate or unrealistic initial conditions. This means that much like induction and abduction, deduction requires information from outside the confines of the logical structure of the argument to assess it. Deciding where, in the argument structure, to derive that information, and how that information is actually of use in assessing different explanations, is a relatively little discussed area in physical geography.

This is not to say that it is not considered and considered in some depth. The importance of sedimentary profiles for reconstructing past environments is of vital importance in Quaternary studies, but why this type of information is used and how it intervenes in the explanatory structure is not usually made explicit. Datable sedimentary markers such as molluscs are vital because they provide a temporal framework for defining when events occurred. Likewise, pollen from sediment can be used to infer the nature of past environments, providing information on the general nature of the environment at different times. Event-specific markers such as volcanic ash could be used to infer if a specific event happened and when. Combined, this type of information can be used to test different arguments about the occurrence of specific events at specific locations at specific times. The type of evidence available may, however, focus an investigator solely on the type of events that can be identified and so limit explanation to these events alone.

In order to try to understand the importance of explanatory, or rather causal, structures in explanation in physical geography it might be useful to rethink the above discussions of different types of explanation as causal networks. Cause and law are known and the effect is a logical and necessary outcome of the two. In induction, only the cause and effect are known. A link could be made between the two, but its status is indeterminate and the number of possible links is high, if not infinite. In abduction, the effect and law are known, but the link is not stable nor is the status of cause fixed. The situation is rarely as simple as is illustrated in these three figures. Often there are a number of intervening states between the initial state and the resulting state.

For deductive explanation, this is not a major problem as long as each link can be viewed as a logically necessary outcome from the previous set of conditions. The states are linked together by the operation of the law. As the number of intermediate states increases, it may become increasingly difficult to sustain the logical necessity of the argument. Uncertainty over what the states are, over the influence of other confounding factors and the possibility of 'other' outcomes, can all affect the confidence a researcher may have in their explanatory structure. This may partly explain the concern with maintaining tight experimental control and keeping causal chains short, as this reduces the need to consider 'complicating' factors. Indeed, it could be argued that short and certain explanatory chains between initial, intermediate and final states are only possible under very specific conditions, and in relation to very specific questions. The movement for process-based geomorphology could be seen as an attempt to ensure that explanatory frameworks were based on short deductively argued chains. Although such studies might derive important general relationships between a specific set of initial conditions and a resulting state under the action of a single law, once the relationship was moved beyond the confines of its controlled system, what happened to the nature of the relationship is likely to become unclear.

Induction seems to have severe limitations as well. In the absence of other information, outside of the confines of the immediate statement of a relationship between cause and effect, the route between the two is almost completely indeterminate. There are a multitude of links that may be possible, that could be argued for, and there is no way of judging them. Similarly, a multitude of intermediate states can be suggested and a range of convoluted linkages put forward to enable even seemingly absurd explanations. As with deduction, maintaining a simple set of links may seem to make an argument more reasonable. Once there is the possibility of a range of possible links, the certainty of any single link is called into question. The effect is known and, supposedly, so is the law, the link between the effect and cause. Working back along this chain, the cause can be found. Abduction is based on being able to tell a plausible story to link effect and cause together via a valid law. There are, however, potentially a range of laws that could be applied to explain any effect. By extension this means that there are a range of potential starting points, a range of potential causes by which laws operate to produce effects. This sounds much like the problem associated with induction, the problems of how do you pin down the law and cause. Unlike induction, however, abduction starts with both law and effect; explanation, in other words, assumes that the law is the correct one to start with.

The issue then becomes forming appropriate links back to the cause. Linkages could be simple, such as a single one, or more convoluted, such as requiring a large number of linkages and the operation of co-varying laws. At this point Goodman's discussion of simplicity (1958, 1967) in such reasoning being the important guide could be invoked to aid explanation. Put simply, Goodman stated that given alternative explanations, preference should be given to the simplest. Defining simplicity is not a simple task in itself, however. Simplest does not just refer to how many links there are in a causal chain, but also to the ease of understanding, the relation of the explanation to other explanations of similar phenomena and the context within which the explanation is to operate. A long causal chain or even causal web may be appropriate in some cases where a particular law provides a simple explanation in terms of always producing a given effect, but which requires a long series of events to link cause and effect. In other words, simplicity is not an absolute property of an explanation but a contextual one. Viewing abductive explanation as being contextually bounded is important. It highlights the conditional nature of this form of explanation. Turner (2004) points out that there are two types of abductive arguments. The first is for unobservable tiny entities, the second for unobservable past entities.

Unobservable tiny entities, such as electrons, are only known through their interaction with other phenomena in experimental situations. Their existence is deduced from their effects. Their use in explanation is that they function as a unifier for a range of phenomena that would be difficult to explain without them. Additionally, they can be manipulated to produce new phenomena through which new properties about them can be deduced. Unobservable past entities cannot produce new phenomena as they cannot be manipulated. Historical entities instead serve as unifiers of phenomena only. The reasons for believing a historical entity indicates that a particular event happened, or that the entity behaved in a certain manner, and may be based on manipulation of or analogy with contemporary entities. Belief, for example, that a specific species of mollusc indicates a change from fresh- to salt-water conditions may be based on observations and manipulation of current mollusc species. This does not negate the fact that the fossil mollusc cannot be directly observed or manipulated. The presence of the fossil mollusc in a sedimentary profile can, however, serve to indicate a particular environmental condition. The fossil mollusc acts as a unifying entity integrating within it information about the nature of a past environment, even if this information is derived by analogy with current environments.

Abductive argument can be biconditional or conditional. An abductive argument is biconditional if the cause thought to result in the effect can produce only that effect. The usual statement of this is that if, and only if, C then E is true. If and only if a specific cause C occurs then and only then will a specific effect E occur. There is, in other words, a necessity for a specific effect to follow only from a specific cause. In most cases, however, it is more likely that the abductive argument is only conditional. The link between effect and cause implies only a probable inference. The usual statement is along the lines: if C is true then E is true. In other words, a specific cause, C, occurs and a specific effect, E, occurs, but C need not always result in E and E need not only result from C.

There is a varying degree of doubt that C will produce E. Abduction searches for the range of hypothetical conditions that could link effect to cause via suitable laws. The laws should link the premises, the cause and effect, in a manner that ideally produces a biconditional argument. Failing this the conditional argument should be a highly probable or stable one. At its heart abduction requires a researcher who can select. Selection of both what makes a suitable law, selection of causal pathways linking cause and effect, as well as selection of what is the 'simplest'. Abductive arguments cannot be decided without recourse to reality, without interaction to determine which of the potential pathways linking cause and effect are present and which, if any, have or are operating. In reality, as von Engelhardt and Zimmerman (1988) note, science is about the interplay of all three types of explanation: deduction, induction and abduction.

Deduction works from cause to effect, abduction from effect to cause. They work through the diagrams in opposite directions. The key difference is that deduction has a logical necessity about its outcomes, whilst abduction can only infer. Deduction permits only one certain link between cause and effect, whilst abduction admits to an infinity of links – but each link has a different degree of certainty or probability. Inserting deductive arguments into an abductive explanation could provide the basis for limiting the potential pathways linking cause and effect, particularly if induction, in the hypothesis-testing sense outlined by von Engelhardt and Zimmerman, can assess the existence of the logic outcome in reality.

Any study of causality in science is likely to involve developing an abductive framework within which competing or different potential pathways can be subject to deductive arguments and empirical testing. The exact make-up of any explanatory framework, whether it is 50 per cent deductive or more, is largely irrelevant; what is relevant, as critical rationalists have noted, is that any explanatory structure is capable of being tested at some point within its structure. The possibility of failure of a structure, the potential for falsification, remains a central feature of scientific argument, but it needs to be placed within a larger context of the logic of the intermingling explanatory frameworks of which it is a part.

Accepting the above argument, that science, or more specifically a field science such as physical geography, is about constructing explanatory frameworks that are both abductive and deductive, the next question is how can we test these frameworks? How can we decide which causal framework is correct or, rather, which is more appropriate for our needs? In the cases above there is no way of deciding which causal framework is most appropriate. Any information can be interpreted in favour of any framework. The issue can be resolved if there is an intermediate effect in the graph, in this case T. What T provides is the possibility of constructing a set of relationships in which the causal linkage of one factor can be eliminated. This is a simple illustration of what researchers have done for ages in experimental designs. Assessing the influence of one factor whilst keeping others constant is not a new idea, but within the context of causation it highlights the importance of understanding cause by intervention.

Causation is no longer seen as something that can be derived from theory alone; determining causation, deciding between causal pathways, requires intervention. Causal networks highlight that causation is a product of human intervention in a system rather than as a property existing independent of that intervention. Causation, in other words, has no meaning without human intervention. Constructing causal networks also highlights another property of causation. A set of entities and relations can be identified and defined as causal networks of a general type.

Any landslide, for example, will be expected to consist of a set of entities and relations. The occurrence of a particular landslide will involve the activation of specific pathways within the causation network. The former causal network is the general or type level, whilst the specific landslide is the token level (Pearl, 2000). The token level is an instance of the general or type level. The two levels are structurally the same, it is only the specifics of each landslide scenario that differ. The two levels interact and inform each other. More token-level events can help to clarify and refine the type-level structure, whilst the description and explanation of token-level events relies upon the structures developed at the type level.

In this manner the stability of certain causal networks is established as more token-level events confirm the structure of the type level. Once a researcher becomes confident of the nature of the type-level causal structures then prediction and extension of the explanation can take place. This may involve stripping the type-level event down to what the researcher regards as the essential components and modelling likely behaviour based on these. The constant interplay between type-and token-level structures can become ossified if the research community believes that it has clearly established the causal network. In this case measurement of an effect and its explanation may become so unproblematic that the effects are not questioned and form a building block in the development of explanations for other effects.

General and Token-type Landslides

The identification of general types of landslides relies upon an extensive knowledge of specific landslides. Detailed knowledge of the causes and characteristics of specific landslides provides the basis for abstracting information that can be used to define generalities, or rather similarities, between specifics. A set of properties that seem to produce similar types of landslide, and upon which other types of landslide can be differentiated, can be identified. These form the basis for classification schemes such as Varnes (1958). At a simple level the classification schemes divide landslides up on the basis of the material within which failure occurs, the speed of failure and the water content of the landslide.

Using such schemes any landslide can be located within the classification matrix. Classification is not just a neutral activity. Identifying a landslide as a mudflow means that you have identified that it possesses specific properties and, in addition, that it will have other properties that you have not identified and will behave in a certain manner. A mudflow will behave as a highly viscous fluid and would be expected to move in a certain manner and at a certain velocity. The specific characteristics of any individual mudflow may vary and velocities will not be identical, but they will be expected to be within an identifiable range associated with that class. Token level or individual landslides are, however, able to influence the classification scheme. Apart from being the initial starting point for abstracting properties for the original classification scheme, individual landslides can behave in a manner that calls into question the expectations of classification. Petley et al. (2002) identified differences in velocity profiles between landslides undergoing failure along a shear plane and failure by ductile deformation.

Each landslide would have similar characteristics of moisture content, failure speed and fail in the same material. Detailed analysis of their profiles of velocity movement prior to failure highlights that the mechanisms of failure are different. In the case of failure along a shear plane, crack nucleation is followed by crack growth that establishes a failure surface. In ductile deformation the stress is such that crack nucleation is possible, but crack growth is not. This means that no shear plane can develop and the mass fails as a single unit rather than along a plane of weakness. This difference is reflected in the velocity of pre-slide movement. Identification of this difference in mechanisms of failure, based on detailed investigation of specific landslides, could be used to modify the classification scheme, the general level.

SUMMARY

Explanation is a means of reducing uncertainty about reality. Explanation implies that there is an interaction between the researcher and reality, based upon an impulse to know about the physical environment and how it behaves. Deciding what is and what is not an explanation implies that there is a set of criteria for judgement. This suggests that explanation is a communal activity based on rules and conventions. At the heart of explanation is a belief in causal relationship being present between entities in reality. Deriving where causality lies, in the succession of events or in the entities, is an important consideration. If causality resides in entities then they take an active role in determining the nature of reality.

Linking cause and effect is a difficult process and relies upon some underlying theory about how reality operates. This provides the basis for deciding that event A should be connected to event B; indeed, that event A should be the cause of effect B. From this simple relationship a different explanatory framework can be imposed. Deduction assumes that cause and the connecting law or mechanism are known and that the effect can be logically derived from these. Induction assumes that cause and effect are known, and that it is only the law or mechanism that is unknown. Abduction assumes that effect and law or mechanism are known and that a plausible cause can be inferred. Explanation usually involves a combination of all three explanatory frameworks. Deciding on the plausibility of a cause-and-effect relationship will involve the researcher intervening in reality to try to control for factors assumed to be important in the explanation.

5

PROBING REALITY

PROBING AND THE DIALOGUE WITH REALITY

Even when physical geographers have developed a theory about how reality operates they still need to determine if this theory 'works' in the 'field'. The 'field' refers to the arena within which the theory is assessed to see if it operates as expected. This could be the field as understood as the natural environment; it could be the field as understood as a controlled subsection of the environment as in field trials. Laboratory-based and computer-model simulations would also count as the 'field' in this context. 'Work' in this context is assessing if the results of probing reality, its measurement, match the expectations that have been derived from the theory.

The assumption is that the 'field' is the same reality whoever defines it, whoever studies it and whatever instruments or techniques are used to assess reality. Recognizing that the field is defined in different ways is not just an acceptance that different physical geographers are doing different things but that they may be operating within and referring to different worlds. Despite these differences, physical geographers have to assume, that there is a physical reality behind their investigations. This assumption does not, however, mean that their 'fields' of practical research need to be defined the same way or approach each other in their characteristics. Much as noted by Goodman (1960, 1978), researchers are into 'world-making'. Important in such world-making are the purposes and objectives of the research. These begin to define the type of reality, the phenomena, which forms the basis of the researchers' views of reality. These objectives define the type of study to be conducted; that is, whether it is a laboratory-based experiment, a field-based experiment or a field study involving direct monitoring of the environment.

Any probing of reality is a form of intervention. Some researchers (Harrison and Dunham, 1998) suggest that, as in some interpretations of quantum mechanics, reality does not exist until the researcher intervenes. They suggest that reality does not cohere, the wave functions do not collapse, until measured by the researcher. Interpreted in this extreme manner, this means that the state of a glacier, the velocity of a river and the entrainment of particles do not exist or occur until the researcher measures them, until the researcher intervenes into the system. However, coherence of a physical system above the quantum level does not necessarily require a human observer.

The quantum system is only observable under the unreal conditions of highly controlled laboratory conditions. Under these conditions the observer is the only entity that interacts with the quantum system. In the 'real' world, quantum particles interact and the wave function is resolved by these interactions with other entities. In other words, the uncertainties of the quantum level are resolved before the researcher intervenes and measures the system; the wave functions have been collapsed. This means that physical reality is not fuzzy and incoherent, nor in a series of potential multiple states before the researcher measures it. Reality is a resolved, singular whole. This does not mean that the researchers will have absolute knowledge of this singular whole, only that they will be interacting and having a dialogue with an entity that exists independent of them. This means that reality will operate and function as it does, whether humans recognize it or not.

Intervention does, however, create different worlds by creating different dialogues. The first step in this intervention is the framing of reality. Framing of reality involves deciding what variables are important, how these are to be defined and measured and, vitally, defining what is not important. This latter point highlights what is not to be defined and considered important to the objectives of the project. Undertaking laboratory experiments could be viewed as the most extreme form of intervention as the reality made becomes the experiment itself. A laboratory is used to exclude all reality other than the variables thought to be important for the operation of the part of reality under investigation. These variables are excluded because they are not thought to be important for the creation of the phenomenon that is the focus of study. Only those variables and relations thought, by some theory, to be important and involved in the creation of a phenomenon will be used in a laboratory study. Even these variables and relations will be closely controlled and maintained at magnitudes and frequencies that can produce the phenomenon with the equipment available. Laboratory studies require, therefore, a clear identification of the phenomenon to be investigated.

This often involves transferring observations of changes in the 'field', the environment outside of the laboratory, and their interpretation as a significant phenomenon. The laboratory conditions then try to recreate the phenomenon as observed. A key question that needs to be asked is whether the phenomenon being replicated in the laboratory really is the same phenomenon as that observed in the 'field'. Hacking (1983) suggests that it may be the case that the phenomena in experiments only exist in the experiments. The phenomena require the experimental conditions to exist. Rather than replicating the conditions in reality that produced the phenomenon, the experiment produces the only conditions under which a phenomenon can occur. These may be the only conditions in which it can exist or conditions that differ from those that produced it in reality. In addition, there is the problem of equifinality in the production of phenomenon in experiments. The experimenter can never be certain that the output from the experiment, the phenomenon, is only producible by the conditions in the experiment.

There may be other conditions under which the phenomenon may be produced and these may be more reflective of the conditions under which it is produced in the 'field'. This problem of replicating reality, or rather the conditions that produce the phenomenon of interest, is a general problem in any study of physical reality. This problem was particularly acute within salt-weathering experiments in the early 1970s. Initial experiments such as Evans (1970) and Goudie (1974) were designed to illustrate the potential effectiveness of salt as a weathering agent. The experiments were designed to maximize the destructive power of salt on rock. This meant that saturated solutions were used and salt supply was fairly continuous. This ensured a supply of an aggressive weathering agent to the rock. This is a situation unlikely to occur in reality. Within the confines of the experiment, however, this produced the desired effect: the rock broke down relatively rapidly. Given the extreme circumstances, however, it was possible the effect might only exist in the experiment itself rather than reflect the phenomenon thought to be observed in reality. McGreevy (1982) makes the point that a great deal of the variation observed in the nature and rates of salt weathering in laboratory experiments may be the result of variations in experimental design rather than variations in the phenomenon itself.

As the desire to make the experiments more 'realistic' increased, so did the potential variation in experimental designs. Alterations to variables such as salt concentration, salt supply, weathering conditions and sample shape all become potentially important factors in producing different degrees and types of weathering. The phenomena identified in the 'field' became increasingly complicated and difficult to understand as attempts to model 'reality' in the laboratory became more 'realistic'. An important problem with these studies is the type of effect measured. The manner in which weathering was measured tended to be by weight loss or percentage weight loss of a sample. The relationship of this index of change to phenomena observed in the field is unclear at best. Rocks undergoing salt weathering in the field exhibit specific forms of alteration such as flaking, pitting and crumbling.

Loss of material does occur, but visually it is the production of distinctive forms that brings a phenomenon to the attention of an observer. Weight loss is a relatively simple index to measure, although it reflects gross losses and gains to a sample rather than just the effect of salt alone. Focusing on weight change as the index for salt weathering severely restricts both the type of experiments that can be run and the type of questions that can be asked about salt weathering. The researcher will focus only on questions that can be answered by analysing weight change. Similarly, the whole experimental design will be dictated by the need to weigh and reweigh samples. This means that any design that does not involve repeated removal of the sample for weighing would not give results that are comparable to other experiments. Once a technique becomes established it is difficult for other techniques to develop to assess the same phenomenon. This is because the results from the new technique cannot be, except in certain circumstances, calibrated with the established index.

Moving to field-based experiments, such as erosion plots or weathering exposure trials, there is less control over the variables and their variations. Experiments, such as erosion plots, try to mix supposed control with the variability and uncertainty that is present in the reality. In erosion experiments, for example, variables that can be measured and whose spatial variability can be restricted are usually 'controlled'. Identification of variables and their properties is dependent on the ability of the researcher to measure them and to ensure that these properties remain constant or different from each other within experimental plots. The researcher has to be confident that the variable and properties measured do not vary spatially by a magnitude greater than the variability between plots. The researcher is relying on the spatial homogeneity of a variable or property to exist.

Similarly, the researcher is relying upon any internal variability of a variable within a plot to have an impact upon the output measured that is less than the impact of the difference in variables between plots. The researcher is, therefore, assuming that they can determine the variability of reality at a scale relevant for their study and that they can also measure a meaningful output from the reality they have controlled. The researcher is relying on their measurement methods and theories identifying and being sufficiently accurate so as to constrain variation to a level that they deem sufficient for their analysis.

Monitoring of the environment suffers from difficulties as well. Although the researchers may not be trying to control reality, or even ensure a lack of spatial variability, their activity is still interfering with reality. The act of measurement is always an act of intervention. An example of this effect are attempts to measure stream velocities. Inserting a probe into a flow will affect the flow itself. There is a feedback between the measurement instrument and the phenomenon that the researcher is trying to measure. Eliminating the feedback and the impact of intervention is difficult as the researcher cannot know what the phenomenon would have been if the intervention had not occurred. The only way to measure the phenomenon is with the instrument that causes the interference. The possibility of using a different instrument to measure the same phenomenon may be seen as a solution to this problem. This entails a major assumption, however. It assumes that the phenomenon has an existence independent of the measurement instrument.

The above illustrates that the 'field' can be defined and constructed by the researcher. Significantly, the field of one researcher is not necessarily the same as the field of another researcher. Each researcher will have different theories to assess and define different boundaries to the world they want to assess. It could be argued, however, that the development of traditions amongst practising physical geographers means that there will be similarities in the bounding of reality. Clear identification of boundaries, as noted by Richards (1996) and Lane (2001), is vital for the development of intensive research.

Analysis of whether such clear and consistent bounding occurs between researchers is unclear. World-making is a vital part of the research process, but it is often neglected (or rather ignored) by scientists as it implies subjectivity in their world-view. World-making does not negate objectivity but it does mean that researchers need to bear in mind the consistency of their world-making as the basis for their decisions.

Measurement Systems

Measurement of a phenomenon requires the use of instruments to probe reality and impart quantitative information about its variability. Instruments are not selected in a vacuum. Although the use of a particular instrument in a particular type of study may become automatic, this does not mean that the use of the instrument is made without some regard to theory. Theory guides the researcher in terms of what to study in the physical environment and also what to measure.

Translating theory into something that can be assessed in the physical environment means that phenomena are made measurable. As with the field discussed above, it is often assumed that the phenomenon being measured is the same no matter how it is measured. Translations of theory into practice may differ, but these do not alter the nature of the phenomenon being measured. Without this assumption measurement between different individuals made in different places or at different times could not be compared. The assumption does not, however, mean that this viewpoint is correct. Within quantum physics there is a view that the observer and phenomenon cannot be separated. The observer and phenomenon make up a single system, a measurement system. In this context it is not possible to separate the measurement made from the measurement system within which it was made. The measurement and the phenomenon become combined in an unbreakable link in the measurement system. It is impossible to talk of a separate existence for the phenomenon and so also, therefore, to talk of an independent measurement of that phenomenon. This means that within the supposedly objective, hard science of physics, it is accepted that reality and how it is measured form an inseparable whole.

Rhoads and Thorn (1996b), based on the work of Suppe (1977), van Frassen (1987) and Giere (1988), put forward a view that the measurement of information in the environment cannot be divorced from the theory guiding instruments used to measure the phenomenon. They summarized this view within the model-theoretical view (MTV) of scientific research. In this view a theory has associated with it a class of abstract structures: its models. These models are all derived from the theory and form a closely related family, related by the same postulates and laws. The models are intimately linked to their parent theory, but the models also need to be linked to reality in order to assess the theory. The linking is, as noted previously, achieved by the construction of hypotheses. These hypotheses are, however, related to the model, to the representation of the theory, rather than directly to the theory itself.

Previously, however, it has been assumed that there is a link between external to the sensor. Similarly, there is a series of theories and protocols associated with appropriate methods of data collection and appropriate methods of data analysis. The layers are potentially infinite. Combined, the auxiliary influences mean that there is a dynamic and ongoing process of construction involved in defining, collecting and analysing any data set. Rhoads and Thorn (1996b) call the aggregate of such a process a data model. The data model ensures that there is compatibility between the data harvested from reality and the requirements of the model to be assessed. As Rhoads and Thorn note this means that a theoretical model specifies the pattern of data a phenomenon or set of phenomenon should generate under a particular set of idealized conditions and the data model reveals whether this pattern of data is, in fact, present in the data collected from the real world.

This means that the theory explicitly guides the researcher and ensures that the data collected are appropriate for its assessment. Data are not raw material, ready for collection and divorced from theory. Data are defined and constructed by the theory via its theoretical models. The wonder is that the same data can be appropriate for the assessment of different models. This implies that the link to a specific model is not limited to a one-to-one relationship. Aspects of data are relevant to different models, potentially related to different theoretical frameworks. This may imply that data are real, that data do exist independently of theory. Alternatively, this property may reflect the limited manner in which models are perceived, and the limitations on what can be measured in the real world. There may only be a limited number of models that can be derived from a theory that are capable of assessment and of these most may require the same data for assessment.

Similarly, what is appropriate data may be defined very narrowly by a specific field of enquiry. This means that a tradition or habit will prevail in what counts as correct data. It should not be assumed, however, that data are static. The information collected by a single instrument may be capable of conversion into other data, other information, using different analytical procedures. Velocity measurements of a single point of flow may appear to produce no pattern when plotted against time. Their subsequent analysis using a running average may highlight patterns that were not visible before. Could you say that the same data were involved in each case? If you accept that data refer to the single measurement points then the answer is yes, but if you view data as the final information produced then the answer is no. Information is always capable of reinterpretation. Reinterpretation depends upon the development and application of new auxiliary theories and upon the impetus to extract such information based upon a new theory or new model that specifies different data for its assessment.

Rhoads and Thorn (1996b) highlight that the MTV view could be advantageous in physical geography. MTV puts the model rather than the theory at the focus of analysis. This is useful in seeming to mimic what actually happens in physical geography, according to Rhoads and Thorn, but it is also problematic. The model is the thing being assessed and so it is also the thing being rejected or accepted.

The status of the theory from which it is derived is always uncertain. Much as in Lakatos's research programmes, the central core can never be attacked, can never be accepted as a firm or even 'real' foundation for understanding. It is only the protective belt of models, in this case, that are exposed to debate. This does, however, mean that study tends to focus on the level of reality appropriate to the model. Assessment is being made of the model derived from a theory that is likely to use other theories based in physics and chemistry for its construction. These 'deeper' theories do not become the focus for rejection, only the models derived from a theory of which they are building blocks. MTV does not require that models be expressed mathematically. This means that cherished concepts in physical geography that are expressed qualitatively are acceptable as theories. This does not imply anything about what is acceptable as data within a data model, however. Physical geographers may be happy to accept qualitative theories, but may be more reluctant to accept qualitative information in their data models. Criteria of acceptance need not be the same for each part of the MTV process.

Central to the MTV is the role of hypotheses, which many physical geographers have argued are a central aspect of practice. The application of MTV to biology, with its similar field-based phenomenon, is a good illustration of the potential of the approach. Although the MTV has yet to be fully applied in physical geography, it does suggest a potential for developing an alternative, but compatible, framework for investigating reality in a field discipline relative to the 'hard' sciences. Physical geography has tended to 'borrow' heavily from physics and chemistry to justify its methods and practices. The difficulty of applying conditions of control and certainty of causality, however misplaced, to phenomena in the field has proved difficult if not pointless. Physical geographers have tended, whether by design or by intuition, to employ a technique that they feel increases the certainty of their explanations: the principle of triangulation. Although the term is again borrowed, this time from surveying, it does act as a good metaphor for the development of practices for obtaining evidence in physical geography. Within surveying triangulation refers to the technique for locating a single unknown point from different known points. Increasing the number of known points from which a single point is observed should fix that point's absolute location more and more accurately.

There is increasing confidence in the belief that the point is accurately located the greater the number of surveying stations used to determine its location. Triangulation within physical geography metaphorically refers to the use of different methods and different sources of information to assess the same model and, by implication, theory. This does not necessarily mean that different data models are being used. The same data model may permit different types of information to be collected, each valid as a means of assessing the theory. When the different methods or different sources imply that same result, then there is a convergence of evidence. When the different methods all diverge, they imply that there is no agreement about the status of the model; that is, whether it is valid or invalid. Their divergence does cast doubt over the status of the model, and hence theory, increasing the level of uncertainty about its explanatory potential. An interesting situation may arise where there is both convergence and divergence. In this case some methods may imply a similar result whilst others may imply a different but consistent result. In surveying this would point to a systematic error in the locating of a point by certain surveying stations.

Multiple Working Hypotheses

The method of multiple working hypotheses (MWH) has been championed by many authors (e.g. Haines-Young and Petch, 1986; Baker, 1999). An early paper to explore this method was by Gilbert (1896) in relation to what he described as a topographic problem. Gilbert called a hypothesis a 'scientific guess' (1896, p. 1) and viewed the initial guess as the first step an investigator made to connect a group of facts and a cause. Rather than starting with a blank canvas Gilbert saw the hypothesis as linking existing cause and effect. The role of the hypothesis was to provide a means for extracting information that would assess the validity of the link between the two. Gilbert reasoned the link as follows: In other words, he [sic] frames a hypothesis or invents a tentative theory. Then he proceeds to test the hypothesis, and in planning a test he reasons in this way: If the phenomenon was really produced in the hypothetic manner, then it should possess, in addition to features already observed, certain other specific features, and the discovery of these will serve to verify the hypothesis.

The hypothesis suggests new features, new observations which should exist if it is true. The investigator searches for these new features and if they are found the theory is supported; if not, it is rejected. Gilbert states that a theory is accepted as satisfactory if the new features found to support it are numerous and varied. Gilbert does not seem prepared to accept a single new feature as sufficient for accepting a theory. Gilbert states that a series of trials will result in inadequate explanations being set aside until one is found that satisfies all the tests carried out. Unlike Popper, Gilbert views this method as leading to verification of a satisfactory theory rather than non-falsification.

Gilbert then applies his technique to the problem of a crater at Coon Butte in north-eastern Arizona. Gilbert put forward four hypotheses that could explain the crater: explosive formation, meteorite impact, volcanic intrusion and volcanic explosion. Gilbert assesses each hypothesis in turn, using existing evidence in the case of the explosive formation and volcanic intrusion, and then outlining the critical testing he undertook to assess the remaining hypotheses. In each case Gilbert outlines a logical, deductive sequence of events that could have resulted in the formation of the crater under each hypothesized cause. Associated with each sequence was a set of features that should be present if the sequence occurred. Data collection and analysis is then focused on identifying the presence or absence of these features for each hypothesis. This means that data collection and analysis is always done with a clear purpose, never for its own sake. The explosive formation was initially supported by the circular shape of the crater and by the presence of iron, which was thought to have been ejected by the explosion. Analysis of the iron suggested it was of extra-terrestrial origin, so eliminating the explosive hypothesis. The volcanic intrusion hypothesis relied upon the presence of a laccolite under the crater. This intrusive form would have forced the limestone and sandstone strata to deform as a dome. The crater would represent the remains of this updoming. The lack of any volcanic rocks within the crater eliminated this hypothesis.

The last two hypotheses were assessed by Gilbert in the field data himself. He devised two critical tests, highlighting his concern that no single line of evidence be sufficient to verify a hypothesis. First, a detailed topographic survey should indicate if volume of material in the rim matched the volume of material excavated from the crater. Gilbert believed, from experimental work with projectiles, that a meteorite with a volume of 7.5–60 million cubic yards was necessary to produce the crater. Absence of any difference in volumes between the rim and crater would indicate that no such additional mass was present. Second, Gilbert carried out a detailed magnetic survey, reasoning that an iron meteorite should leave a high iron content in the crater that would produce a deflection of a magnetic needle.

The topographic survey found the same volume of material in the rim as excavated from the crater, 80 million cubic yards. Likewise, the magnetic survey found no deflection within the crater relative to the plain upon which it was located. Both these critical tests meant that the meteor hypothesis should be rejected. Although Gilbert does reject the hypothesis, he appears reluctant to do so. In the text he states that this was his favoured hypothesis before fieldwork began. Even when the data confronted him, Gilbert resorted to experimental work to assess if the meteor could have been smaller or at a depth where its magnetic influence would not be detectable at the surface. Although he does not accept the meteor hypothesis, his comments leave it as an uncertain rejection rather than one condemned outright.

The remaining hypothesis is not contradicted by the field data, but neither does it seem to be critically verified. Gilbert argues by analogy with other craters that are thought to have been formed by volcanically induced steam explosions. It is important to recognize that Gilbert's analysis does not just rely on the field data. Gilbert switches between argument by analogy with other known forms and to experimental work on processes such as projectile impacts. He uses the information from each to try to justify his deductive arguments about what features should be expected for each hypothesis. If the links between experimental work or analogy and his argument are later shown to be erroneous then his chain of reasoning breaks down. The results and rejections rely on his reasoning being valid; if it is not, then the status of each hypothesis could alter.

TRIANGULATION OF TECHNIQUES – MEASUREMENT OF SURFACE FORM ON ROCKS

Analysis of the morphology of weathered surfaces has used a variety of techniques. Trudgill et al. (1989, 2001) used a micro-erosion meter to measure changes in the height of points on a surface relative to a reference plane. An indication of change for each surface was derived by averaging the height changes for 120 points over each time period. It was assumed that 120 points provided a satisfactory representation of the surface. Williams et al. (2000) used a laser scanner to characterize surfaces on wave-cut platforms. Their method collects tens of thousands of height points in each time period. Inkpen et al. (2000) used close-range photogrammetry to derive a photogrammetric model of rock surfaces. The number of point data on any particular set of images collected depended upon the requirements of the operator in analytical photogrammetry and the software in digital photogrammetry. Both papers assume that the surface they are trying to measure is real, but both also highlight that the techniques being used have problems, or rather limitations, in providing a 'true' measure of the surface. Any measurement technique carves reality up in a distinct and limited number of ways. The entity resulting from the carving of reality is not understandable outside of the context of its associated measurement technique. Linking entities between techniques implies an ability to agree on the presence of an identical carving of reality.

In both studies, the surface heights are only the final outcome in a complicated network of relations. Hidden behind these heights are a raft of other networks such as technical experts and commercial organizations, as well as material networks such as electronics and other equipment. These networks are often hidden and become 'black boxes' of which the individuals measuring the surfaces know very little. With the laser scanner, for example, detailed knowledge of the electronics and how these influence the operation of the scanner may be required to interpret the behaviour of the laser. This is contained within the software used to analyse the returning laser light. With commercially available scanners, it is likely that few users would understand the equipment in such detail or even be aware of how different contextual conditions may influence results.

An important part of these intertwined networks is the training required to produce the surface heights. Close-range photogrammetry, for example, requires the user to understand surveying techniques, photogrammetric techniques such as the 'floating point' measurement of heights, and to establish appropriate control networks. The implementation of this training requires existing and often expensive capital equipment, such as the total station and analytical plotter. The recent and increasing use of digital photogrammetry to interpret stereo-images does not remove this knowledge, but again hides it within the software. Each paper uses a different property of the surface to determine the nature of that surface. The laser scanner measures surface height by measuring the time it takes for light emitted by the laser to return from a surface.

The return time is a function of the laser velocity as mediated through interaction with the target surface as well as its transmission through the instrument lens and detector. These data can then be related to the control frame to permit a calculation of surface heights. Close-range photogrammetry uses stereo-images of a surface taken with a calibrated camera as the data source. There are particular photographic requirements that must be met to provide an adequate pair of images; for example, at least 60 per cent overlap between the images. The surface properties as detected by the film and transmitted through the lens are the basis of all further manipulation of the information. A long procedure of image interpretation is then required before a product of surface heights can be produced. With digital photogrammetry there is often the additional step of scanning photography to go through, a step which can introduce another source of change into the information contained in the images. These differences in the routes used to determine the same product, surface height, can remain hidden once the data are presented and interpreted as a real set of surface heights.

Despite the similarity in the final output, surface heights, and the commonality of the subject area, both papers are associated with slightly different networks of relations that define their techniques. Both networks, however, provide for stable translations of instrument readings into representations of a 'real' surface. Similarly, both networks have a set of established common practices and are highly dependent on the crystallization of network relations into durable and expensive equipment. Both sets of height values are interpreted as if they refer to the same entity. Outputs from both measurement techniques are assumed to be different pictures of the same real surface.

The differences between the two networks are reconciled through the idea that they refer to the same entity: surface height points. The unproblematic nature of these points hides the differences in training and errors. Differences in surface heights are usually interpreted as errors that displace the surface heights from their 'true' values. Assuming a common external reality removes the point measurement from the measurement system of which it is a part. Taking the point measurement out of this context means that it can be manipulated and interpreted independently of its origin and, significantly, context. It might be argued that the various control networks and systematic testing of each technique means that the results produced are reliable and do refer to the same underlying reality.

Likewise, comparison of the two systems by measuring a standard surface should illustrate their accuracies and errors. Such testing and control is, however, only ever made within the context of the measurement system itself. How can a standard surface be produced without a measurement system to define it as a standard surface? The techniques measuring the surface would still produce error terms specific to and an integral part of that unique measurement system. Although it would be assumed that both techniques provided the same data on the standard surface it could equally be argued that each measurement system is measuring a property of the surface that has been defined as surface height. The surface and instrument properties each measurement system is using to derive information could be different, but within our interpretative network they are lumped together as surface height.

It is the assumption of real external surfaces with a singular property called surface height that permits both the accommodation of the information from each technique into a common interpretative framework and the decontextualization of the information about surface heights be made for possible reasons why these stations produce such systematic variation. A similar conclusion might be drawn in physical geography. The presence of both convergence and divergence may point to a systematic variation in what each model, and the associated data model, explains or indicates about the phenomenon. Such differences may point to different theories and models illuminating different aspects of a phenomenon rather than being the explanation of a phenomenon.

PRACTICE IN PHYSICAL GEOGRAPHY

Putting any theory and its associated models and data models into practice, actually assessing them, involves the development – or rather construction – of a location for that purpose. Location can refer to a laboratory as much as to the field or experimental plot. The important point is that all locations are prepared in some way; they are not reality as it is, they represent reality as conditioned and constructed, often physically, by the researcher. This construction has been noted above, but the question becomes how is this carried out in any study? Richards et al. (1997) and Richards (1996) provide some indication of the approaches available to physical geography. Richards et al. (1997) view two relationships as key in measurement: from the real to the actual and from the actual to the empirical. In other words, taking a realist perspective on measurement, the key relationships concern the interpretation or mediation between the levels of reality, the real, the actual and the empirical. Richards et al. (1997) see physical geography as having vague and imprecise concepts such as equilibrium, and it could be claimed more recently self-organization, that are translated into entities capable of measurement, as noted in Chapter 2.

They see a problem, however, in this imprecision. Imprecision means that there may be more than one translation thought to be acceptable, and so different, distinct entities are classed as the same thing. Equifinality appears because of this imprecise translation. The second relationship, between entities and facts, they term 'transduction'. Relating their definition to Pawson (1989), they view transduction as the conversion of an entity to a measurable quantity. The process involves conversion of one thing into another, or rather the representation of one thing by another. Different measurement systems could be used to identify and measure an entity and all could concur that it exists and responds in a similar manner to each measurement system. In this case it could be argued that there is increasingly greater confidence in the 'real' existence of the entity.

The idea is, as noted by Hacking (1983) as well, that if different measurement systems with different auxiliary theories identify and quantify the same thing in the same manner then there is likely to be something 'real' behind the measurements. Whilst this is a reasonable starting point, the researcher can never be certain that what the measurement system coincides or triangulates upon is the entity that they believe exists. It may be an entity, but its properties may produce effects beyond those measured and its actions may only be measurable through its constant co-occurrence with other entities in the measurement contexts studied. These problems may become clearer if fieldwork is undertaken. This is another case of the importance of practice and theory informing each other. Richards (1996), building on previous work, sets out the case for different types of research strategies within physical geography. He tends to build upon the intensive/ extensive division suggested by Sayer (1992) for use within the social sciences. In this framework, an extensive research strategy, which Richards broadly equates with more 'positivistic' science, identifies, defines and even quantifies relationships between entities by using a large number of measurements or samples.

These are samples of reality. Samples are used in extensive research strategies to illustrate the product or the form of a variable. Samples are required in large numbers because they are used to define and quantify statistical relationships. More samples means greater confidence in the presence and nature of the statistical relationship. Extensive research strategies focus on gathering information on generalities, on patterns of forms. This requires a lot of information on these forms, and so requires a large number of samples. The samples studied possess the property in which the extensive research strategy is interested. Individually samples may possess other properties that could vary between samples.

Unless these other properties dramatically disturb the predicted pattern produced, they are invisible in the extensive research design. Extensive research is not interested in assessing the complexity or even complicated nature of the production of a particular form. Richards (1996) suggests that this reflects the usefulness of the extensive approach for positivistic and reductionist views of science. Patterns exist across different specific contexts. Data are collected concerning these patterns from different specific contexts and so they represent generalities that can be discovered in reality. The relationships that produce these patterns can be quantified and even modelled. Predictable outcomes from simple interactions of specific variables are the end result of extensive research.

Why these patterns occur, and why they occur in specific situations and not others, is not really within the scope of an extensive research design. Intensive research strategies are, according to Richards (1996), more in tune with a realist perspective on reality. Examples rather than samples are the focus of this strategy. Intensive studies deal with few or even only one detailed case study. The purpose of this case study is to identify and explore how mechanisms (the real) operate in a given set of circumstances (the actual) to produce measurable alterations (the empirical). Case studies permit analysis of phenomena across all three levels of realism's stratified reality. Central to the development of the case-study approach is a clear understanding of the theoretical aspects of a location.

Abstraction of a theoretical basis for entities and interactions within a location provide the basis for any case study. From this framework, expectations of behaviour can be derived and applied to help in measurement of the environment. Observations within the case study are the contingent outcome of the presence of general mechanisms and the coincidence of appropriate entities for the mediation of them. These entities exist only within a specific spatial and temporal context: the field location. Their mediation of mechanisms produces the range of measurable variables used in analysis. Mechanisms operate with differing intensities because of the uniqueness of the context, the uniqueness of the entity mix, with some mixes enhancing the action of a mechanism whilst others inhibit it totally. It is only through an exploration of these different and unique contexts that the generality of mechanisms can be recognized and, importantly, the conditions of their operation even partially understood. This form of strategy is viewed as more appropriately suited to trying to understand detailed process–form relationships.

Case-study protocols have been commonly used within social sciences. Their role is much as identified by Richards (1996): the application of theoretical structures to specific situations or contexts. As in Richards's scenario, this application permits a greater understanding of the operation of a theoretical structure by identifying the parameters and limits of its actions. This strategy, however, relies upon the appropriate identification of a theoretical structure and its translation into a field location. Richards (1996) suggests that rather than field studies just being either extensive or intensive, it is more appropriate to view them as continually switching from one form to the other: However, these two styles of research, and the generalisations and explanations that they generate, are inextricably linked, and in any area of geomorphological enquiry there is a continual spiralling between them. This reflects a movement of the study from outside the case(s) in more extensive investigations, to inside a case in an intensive investigation (when new questions may be posed that require the embedding of additional extensive enquiries within the intensive case study).

The important point is that the two forms of strategies need to be combined to study physical reality. Constructing a theoretical framework for intensive study requires some initial starting point, often provided by information derived from extensive studies. Extensive studies can help to identify entities that may be of significance within a particular theory, that tend, for example, to co-vary in a predictable manner producing distinct patterns. By using this information it is possible to identify field locations where these variables are present and where their operation should be open to monitoring and further analysis. Vital to this latter task is the appropriate closure of the location. This means that the identification of the system under study and its environment are significant preliminary stages of any investigation. Identification of boundaries and the theoretical context within which they are identified are important aspects of any investigation.

Likewise, the measurement of a range of entities and their values within the bounded location is important. Rather than just a single property, intensive study is interested in characterizing the location. For this it is necessary to understand the detailed spatial and temporal variations in the location. The operation of a mechanism is likely to be reflected in and impact upon these variations in process and form. The purpose of the intensive study is to trace the mechanism in its mediation to the actual and empirical. This means detailed and wide-ranging measurement of entities and properties rather than just a simple single property measurement.

The analysis of the operation of mechanisms can only be made if the conditions of their operation are fully understood. Although this may be impossible to achieve practically, the narrowing of the potential uncertainties in the system can help to narrow the potential range of actions of mechanisms. It is important that the properties of any selected field location are outlined clearly and justified in terms of the mechanisms under investigation. In this manner it becomes clearer why specific results might be expected at these locations. Establishing the action of mechanisms at one location does not, however, imply that these mechanisms operate in a similar manner elsewhere.

This is where further extensive studies to identify locations with similar properties and with dissimilar properties are necessary. Intensive study of similar locations provides information on the general nature of the operation of mechanisms for a given set of conditions and how variation of these conditions affects the mechanisms. The dissimilar locations can help to assess if mechanisms can operate even when specific conditions appear to be inhibiting. Additionally, undertaking an intensive study may require the use of extensive surveying techniques to analyse a particular aspect of the operation of mechanisms in a location. For example, in a detailed study of velocity structures within a stretch of river, intensive strategies involving monitoring of a range of parameters may still require the use of a sampling strategy to select appropriate points. The points sampled are then likely to require further statistical manipulation to produce a 'representative' numerical description of the flow. Although the statistical manipulation may be tightly constrained by established practices, the data collected and analysed still require both extensive and intensive methods of data collection.

LINKING PROCESS AND FORM – INTENSIVE STUDY OF BEDFORMS

Interpretation of process and form links in physical geography has often been viewed as a simple cause-and-effect relationship. Process is equated with cause and form with effect. The relationship is likely to be more complicated than this simple view. Lane and Richards (1997) assessed the linking of form and process within a specific river channel. Initial analysis of the data from an active braided reach of a river, the Arolla, in a glacierized catchment in the Swiss Alps, suggested that there was a good correlation between the magnitude and direction of a change in discharge and the volume of material eroded or deposited.

Large events produced the highest erosion. This observation would support a simple interpretation of magnitude and frequency relationships outlined by Wolman and Miller (1960). Closer analysis of the data suggested a more complicated relationship, however. Sediment supply seemed to also influence the magnitude of erosion or deposition.

Specifically, points later on the rising limb of the hydrograph showed less erosion and even deposition in some cases, whilst the falling limb often showed deposition independent of the magnitude of flow. This suggested that both discharge and sediment supply controlled channel change. The situation was even more complicated. Sediment supply itself was determined by patterns of upstream erosion and deposition as well as local supply of sediment from bank erosion. Upstream sediment supply depends on the interaction of daily discharge with available sediment.

The possible multiple combinations of discharge and sediment supply mean that it is impossible to specify in advance what discharge will be the 'dominant' flow within the channel producing alteration. The concept of a 'dominant' or formative event can only be meaningfully understood within the context of the whole catchment. Understanding a particular reach where measurements need to be taken cannot be divorced from the relationship of that reach to the rest of the catchment. This view of magnitude and frequency relationships also meant that the response of a reach to an event could not be understood without reference to the history of that reach, the 'conditioning' effect of previous events. Lane and Richards suggest that history has a spatial manifestation in the morphology of the reach. Process of sediment transport will be influenced by channel morphology and vice versa. As the channel morphology alters so the sediment-transporting capacity of the channel will alter.

Initial deposition on both sides of the channel, with erosion confined to the scouring of a narrow channel, produces a morphology that will influence further zones of deposition and erosion. The lateral depositions acted as sediment sources for the next morning flow. Consequent downstream deposition encouraged the development of a new medial bar. The sensitivity of different areas of the morphology to change varied as well. The fossilization or otherwise of features depends upon their interaction with process events, that are themselves mediated by morphology. Lane and Richards suggested that the long-term development of the main medial bar was determined by the effects of processes operating at a shorter time-scale and smaller spatial scale. These processes, however, altered as the morphology evolved. This means that process events cause change in the morphology of the reach, but in turn the process events are mediated by the initial morphology. The precise response of a reach to a process event will vary as the morphology evolves. History, as noted by Schumm (1991), is important for understanding physical systems.

Probing Reality – Fluvial Flow Structures

The nature of river flow has recently been the subject of debate in the study of gravel-bedded rivers (e.g. Yalin, 1992; Biron et al., 1996; Bradbrook et al., 1998; Lane et al., 1999). Buffin-Belanger et al. (2000) set out to confirm the existence of large-scale flow structures and develop a technique for improving the visualization of these forms. From this technique, they suggest that the high-speed 'wedges' of flow display a complex organization. This paper illustrates a number of points relevant to this chapter. The construction of the entity of study – the high-speed wedges – relies upon the use of specific detection methods and a specific theoretical framework. The new technique of analysis actively constructs the new entity of analysis, the velocity fluctuations in a space–time matrix. Lastly, the outcome of their analysis provides a basis for further discussion of their constructed entities. The entities can be treated as if they were real and so debated amongst other researchers in the field. This does not mean to say that they are 'real' or understandable isolated from the measurement system, only that they can be treated as such for discussion. Buffin-Belanger et al. (2000) begin by noting that Falco (1977) modelled a link between small-scale eddies and larger-scale flow structures.

These forms could be identified by their velocities relative to the velocity of the flow as a whole; they were either higher or lower. Similar theoretical and empirical structures were identified by Brown and Thomas (1977) and Nakagawa and Nezu (1981). Importantly, these large-scale features have been identified by a combination of procedures involving multiple sensor arrays and visualization techniques rather than single sensors. The flow structures are only identified when time periods are aggregated and compared to the average flow velocity of the river. The flow structure is an artefact of a comparative computational process. This highlights that the entities are represented only within computational analysis; without this computation, and decisions about periods of time being above or below the average values, the entities would not exist within studies.

There has been a triangulation of different techniques, upon what is perceived to be a real set of entities, that seems to increase confidence in their reality. Kirkbride and Ferguson (1995) and Ferguson et al. (1996), for example, use Markov chain analysis on velocity data from three different heights in a natural gravel-bedded river to identify different velocity states. Roy et al. (1996) use a single-probe burst technique to identify flow structures near-bed and in the outer reaches of flow. The flow structures are assumed to be real and so must possess certain properties, common to all flow structures, which can be measured. It is these properties that the measurement systems are designed to identify and quantify. The runs of higher and lower values than average have a pattern that reflects the theoretical structure of the entity, of the flow structures. The convergence of different measurement techniques on the presence of these runs of values is viewed as evidence for the reality of these entities. The measurement systems used, however, are all designed to identify the units that make up the entities, the velocity measurements of a specific time slice.

Buffin-Belanger et al. (2000) measure their velocities in a straight section of the Eaton North river, Quebec, Canada, at relatively low flow. The river had a poorly sorted gravel-bed with a D50 of 33 mm and a sorting coefficient of 2.9. Mean flow depth was 0.35–0.40 m and mean flow velocity was 0.36 ms–1. A beam was stretched across the river and a movable support arm attached to this. Three March McBirney bi-directional electromagnetic current meters (ECMs) were fixed 8 cm apart on a wading rod attached to the support arm. Each probe had a sensing volume of 3.25–3.9 cm^3 and a low-pass RC filter with a time constant of 0.05 s.

These sensors were arranged to permit three simultaneous measurements of streamwise and vertical velocity components. The simultaneous measurement of these components was essential for the latter processing of the information to derive representations of the flow structures. The data model even at this stage influences the nature of the measurements taken. Twenty-three combinations of three simultaneous velocity measurements along five vertical profiles were taken along a 1.3 m transect parallel to the average flow streamline. This means that the latter condition of the average flow streamline had to be identified in the field and the equipment moved to be parallel to it. This restricted the experiment to a relatively straight stretch of river, as already noted. This could restrict the phenomenon being identified to only these experimental conditions rather than identification of a universal feature of all stretches.

In each vertical profile initial velocity measurements were taken at 3.5, 11.5 and 19.5 cm, with additional profiles being taken at 2 cm intervals up the rod until a height of either 25.5 or 27.5 cm was reached (the water surface). This experimental design produced 12–13 elevations with a total of 69 points of velocity measurements. At all points the streamwise (U) and vertical (V) velocity components were sampled for 60 s at a frequency of 20 Hz. In addition to these data additional information was obtained from a triplet of measurements over a 20-minute period at a frequency of 20 Hz. This permitted an assessment of the temporal variability of flow structures identified.

The above detailed description of the location and method provides information on the requirements for identifying the phenomena. The entities trying to be identified are not obvious from visual examination of the river. Instead, the entities are believed to exist from theoretical considerations, but their presence can only be confirmed by their signal in a specific and well-controlled set of information. This means that a detailed and well-confined experimental design is required to generate the quantity of information that can be statistically processed to derive appropriate representations of the entities. Do the phenomena exist outside of these carefully constructed experiments? For discussion of the entities, it has to be assumed that they do, but argument about the nature of these entities requires the same painstaking control of the environment to glimpse the entities.

Analysis of the data takes a similar approach of triangulation. Many statistical interpretations of the data are used, rather than a single one, to obtain different representations of the entities. Space–time correlation analysis (STCA), based on the calculation of correlation values between samples at different spatial locations and at different time lags, is used. From STCA the maximum correlation values (rmax) and the time lag of this, (Lmax), are key derived properties of flow.

A multi-signal detection technique is also used, based on computing the joint time-series by averaging instantaneous velocity fluctuations for the three signals over a time period (1 s in this study). Flow structures are detected using the U-level technique (Bogard and Tiederman, 1986). This is based on the magnitude of the deviation of an instantaneous velocity value (standardized) from the average value of the time-series. High- or low-flow velocities are identified relative to some threshold value, in this case the mean and variation from it. The key innovation is, however, the use of a novel method for visualizing flow structures. Velocity fluctuations (deviations of instantaneous velocity from the average of a time-series) are represented for each height as either a positive (black line) or negative (white line) deviation.

The resulting matrix provides a visual repre-sentation of the persistence or otherwise of flow structures. Buffin-Belanger et al. suggest that the matrix is like an unrolled film beginning at the right edge and ending at the left. Visually it seems similar to a bar code and could be seen as representing the 'signature' of a river's flow over a given time period. With this method large areas of black or white represent the persistence of flow structures, and the timing of these structures or wedges between the sensors can be determined so identifying the characteristics of the shape of these flow structures. Refinements can be made to this matrix by adding thresholds for low and high flow enabling 'noise', small fluctuations, to be downplayed relative to the large fluctuations at the core of the high- and low-velocity wedges.

The whole paper is concerned with being able to identify and visualize the flow structures. The structures as such cannot be sensed directly, but what can be sensed are the expected properties of these entities. Changes in velocity are a representation of these entities, although they are only a single, specific property associated with them. To this single measurable property a range of statistical techniques have been applied that seem to converge on both the identification of the entities and on their nature. Whether the convergence is because the entities are real or because the measurement systems are searching for the same type of change is irrelevant. The information is treated as if the entities are real and have a representation in the data. Flow structures are constructed by both sensors and statistics, but their reality is no less 'real' because of this. Their reality lies in their use as objects for discussion and analysis, objects that can be identified by others with measurement systems and their nature contested and tested by others.

Computer Simulation of Reality

Computer simulation of reality is another means by which the operation of reality can be probed. Models are particularly relevant within physical geography where they can provide information on changes that may be beyond the lifespan of investigator or project funds. By building process-based models, for example, it may be possible to link process studies to long-term evolution of the landscape. Likewise, models can provide useful 'what if' scenarios where manipulation of reality at a scale sufficient to provide such information is not possible. Haines-Young and Petch (1986) class simulations such as the global circulation models (GCMs) used in climatic-change studies as Aristotelian or pure thought experiments. They produce results, but are the results testable? For GCMs the time required to wait for testing of their outputs is too long to be practical. It could be argued that models such as those used to simulate hydrological properties of rivers do provide testable results.

These models can be assessed as any other hypothesis concerning reality. An unfalsified model does not, however, mean that the model provides an appropriate simulation of reality. A model may produce a result that cannot be falsified, but one that also does not reflect how reality operates. A model that produces patterns that look similar to those found in reality does not necessarily imply a model that accurately reflects reality. Haines-Young and Petch (1986) argue that models are often indistinguishable from theories and hypotheses in much of scientific writing. They focus on the role of models as devices that can make predictions. Models take existing information and process it, based on a set of rules. Processing this information produces new information that was not immediately known from the original information. Models provide a formal means of deriving hypotheses about reality.

Beck et al. (1993) argue that a model is a complex assemblage of constituent hypotheses. A model is based on a set of hypotheses that operate together to produce a set of outcomes that can then be compared with reality. They also suggest that there are three objectives to modelling: to predict the future, to identify the mechanisms crucial to the generation of patterns observed in reality, and to reconcile observed patterns with concepts via the model. They view the model as an essential tool in simulating reality.

The simulation becomes the focus of study and, in particular, its relationship to observed patterns of phenomena. Beck (1991) outlines a threefold typology of models: metric, conceptual and physics-based. Metric models are based upon observations and try to characterize system response from these observed data. Observations are the driving force in developing these models, with empirical data providing the raw input into the model. Conceptual models identify the model structure before any data are collected. These models already have the important processes identified and use environmental data to calibrate these processes. Physics-based models work from first principles and build a physically coherent model of reality from these building blocks. Kirkby (1996) argues that the preferred models should be physically based. In other words, the most appropriate models are those that have some physical mechanism or relationship as the basis of their construction. The model applies and extrapolates the operation of this mechanism to a specific set of parameters to provide an outcome that can be compared with reality.

Although a model may be seen as a simplification of reality, this itself is too simple a view. A model summarizes what an investigator regards as important about the section of reality under study and how it operates. Parameters and relationships used in the construction of a model are selected by the investigator, hence they reflect their a priori view of what is important and how entities of interest are related. Kirkby (1996) suggests that effective models should be simple. This means that they should specify reality in as few terms as possible.

Models should be sufficiently general in their specification to allow transfer between areas, but sufficiently rich in their structure to allow diverse work within the subject area to be understood via their operation. The model provides, more than a quantification of reality, a means of thinking about reality. Models provide a means of assessing if the understanding we have is sufficient to provide an adequate explanation at other levels. For example, does our understanding of processes at a micro-level provide a sufficient basis for extrapolation to landscape-scale processes and forms? Any model cannot, however, prove the validity of our processes as explanations. Models can only provide a possible explanation that does not contradict existing knowledge, or rather one that is consistent with it.

SUMMARY

Dialogue with reality requires analysis in the field. The 'field' can be defined in different ways and reflects, to some extent, the different ways in which different physical geographers construct the reality they are studying. Dialogue requires intervention. Intervention means a researcher has to frame reality, has to decide what to include and what to exclude from their consideration. Examination of reality requires the use of measurement systems to provide information about the nature of phenomena and their variations. All measurement will be theory-laden, but this does not automatically mean that all measurement is subjective. The model-theoretical view (MTV) illustrates the intimate link between theory, data models and hypotheses. Theory guides the researcher as to what data to collect, ensuring that the data are compatible with the identification and quantification of the type of phenomena the theory expects to be present. Triangulation through different instruments upon the same phenomenon can increase the confidence researchers have in the reality of that phenomenon. In such cases, the basis of each measurement system needs to be considered in detail as they may be based on the same principles rather than reflect different data models.

Increasingly, the complexity of the physical environment is being recognized as extensive studies of entities and relationships yield little in the way of explanation. Extensive studies focus upon statistical relationships and the identification of generalities and patterns through the use of a large number of contextless samples. Intensive studies put entities and relations back into their context. These types of studies are concerned with why patterns occur in specific situations and not in others. Intensive studies identify and explore how mechanisms operate in a particular set of circumstances to produce empirically measurable change.

6

SYSTEMS: THE FRAMEWORK FOR PHYSICAL GEOGRAPHY

SYSTEMS ANALYSIS IN PHYSICAL GEOGRAPHY

Systems analysis came to the fore in physical geography in the 1960s and found formal expression in the classic textbook of Chorley and Kennedy in 1971. The introduction of systems analysis into physical geography was not without critics, but the new approach rapidly became one of the cornerstones of thinking about the physical environment. The continued success of systems analysis is probably best illustrated by a perusal of any set of modern textbooks on introductory physical geography. Often the titles of such texts use 'systems' as an explicit indication of their approach. Almost invariably the contents pages divide the subject matter into specific environmental systems, each of which is considered in turn – for example, the atmospheric system, the lithospheric system, the biosphere, and occasionally with a chapter attempting to integrate the disparate systems at the end of the text. Systems analysis, if judged by column inches of text in undergraduate books and even research papers, has become the overarching framework for understanding in physical geography. With such success it is important to understand what systems analysis is. Systems analysis has also moulded thinking about the physical environment. Entities and relations are viewed in a specific framework and studied according to expected modes of system behaviour. An understanding of the constraints on thinking imposed by systems analysis needs to be clear, so that the limits of systems analysis do not become a barrier to comprehension of the physical environment.

Systems analysis did not develop in physical geography – its pedigree is far longer. 'Systems', as a term, has been around since modern science developed in the seventeenth century. The rise of systems analysis owes a great deal to the attempt to develop an integrated and all-encompassing framework for all science in the twentieth century. The existence of such a framework implies, first, that all reality is capable of being understood; there are no areas or topics outside of its analytical scope. Second, all reality can be understood in a common framework using the same sets of terms. This means that understanding in supposedly different subject areas does not require specialist terms or specialist knowledge, but rather translation of these terms to the common terminology of systems analysis. Third, as there is a common framework, all reality can be expected to behave as predicted by this framework. All reality becomes potentially predictable and, by implication, potentially controllable. Systems analysis should not, however, be viewed as something that emerged fully formed for incorporation into physical geography. Similarly, systems analysis has not remained a static form of study in the 40 or so years it has been used within physical geography.

Having said this, the basic tenets of systems analysis remain pretty much the same as they did 40 years ago and the implications of the approach for how the physical environment is studied have become ingrained within geographical practice. Van Bertalanffy proposed general systems theory as a unifying framework in the late 1930s. Tansley pushed the concept of the ecosystem in the same period. Both had a similar view of a system as an integrating concept unifying the different entities found in an environment within a common analytical framework.

The advent of cybernetics in the 1940s and 1950s added another level of understanding to the systems approach (Shannon and Weaver, 1949). Cybernetics is an underrated influence on the development of systems analysis within physical geography. Physical geography tends to focus on its perceived antecedents within ecology, biology and physics as the source of its development of systems analysis. As such these subject areas hark back to the pre-war era of geographic research. The work of Gilbert (1877, 1896) or Penck (1924), for example, is reinterpreted in system terminology. Gilbert's use of terms such as 'system', and his systematic approach to analysis, is taken as an indication of systems thinking even if the terminology had not been invented. Likewise, Penck's view of landscape development is often recast as a systems diagram (Thorn, 1988) even though the term was never used by him. Both examples reflect, as noted in Chapter 1, a tendency to reinterpret the past in the light of current thinking. The influence of these individuals and subject areas is not denied, but the nature of the systems they suggested was not the same as 'systems' as used and understood today.

Cybernetics, the study of self-regulating mechanisms in technology, along with the development of information theory, had a profound influence on the nature of systems analysis that physical geography encountered in the 1960s. Shannon and Weaver's development of a mathematical basis to information flow and interpretation provided a means of quantifying change in abstract phenomena. Coupled with the lexicon for describing relationships between entities provided by cybernetics, these new analytical frameworks resulted in the development of a highly mathematical and formalized description of systems analysis by the 1960s. It was this version of systems analysis that physical geography tried to bring into the heart of the subject. The success of systems thinking cannot be denied within physical geography; the total incorporation of a fully fledged systems analysis is harder to justify.

The key ingredients of a system are the variables or elements, the relationships between the variables or elements, and the bounding of these variables and relationships from the rest of the world. Hall and Fagan's (1956) definition of a system has been used by a number of authors as a starting point (e.g. Thorn, 1988; Huggett, 1980). Their definition is: 'A system is a set of objects together with relationships between the objects and between their attributes' (Hall and Fagan, 1956, p. 18). The importance of the previous chapters for understanding the limitations of systems thinking should be clearer now. The definition of variables and relationships implies an ability to define and divide the world into distinct entities and relations. Likewise, the definition and bounding or closure of the system itself requires a particular view of reality as divisible and understandable by this division.

Systems thinking depends on the reality of the physical environment being displaced from the observer. The observer defines a system made of real entities and relations, the sum of which becomes a sort of super entity with its own properties and relations to the rest of the physical environment. This super entity may or may not be simply the sum of its elements and relations. The observer is a passive and objective interpreter of the system, almost by definition outside of the boundaries they have imposed.

A distinction is often made between a system and a model (Thorn, 1988). A system is viewed as an abstraction that is assumed to exist in reality. A model is a fully specified, although abstract and incomplete, version of reality. A model is an abstraction and a simplification of reality and it is recognized that it does not, nor is intended to, mirror reality. The distinction is made to clarify the purpose-led construction of models as opposed to the supposed universal nature of systems. A model is designed to serve a purpose. It does not need to specify reality fully, nor to be agreed by all. A system may be unknowable in full, but agreement can be reached that such a set of entities and relationships exists. The distinction may be helpful in distinguishing an operational model from an abstract, but universal, system, but it has the side effect of implying that the system is in some manner more 'real' than the model. The system has some universal status whilst the model does not. Systems, therefore, seem to be a closer representation of reality despite the user-defined nature of all their entities and relationships. Within physical geography it is usually system models (Thorn, 1988) that are being considered rather than systems as universal abstractions.

APPLICATION OF SYSTEMS THINKING

Development of typologies of systems application has resulted in long lists of different types of systems based on a range of criteria. Chorley and Kennedy (1971) provide one of the first based on the form and function and complexity of the systems studied. Complexity increases from simple cascading and morphological systems to process– response systems and on to biological and social systems. Strahler (1980) develops a typology based on similar criteria to Chorley and Kennedy. Terjung (1976) uses four criteria to separate system modelling in physical geography into different levels. The criteria used relate to the type of logical argument used in explanation (induction or deduction), the level of explanation (individual entities or the system as a whole), the degree of deterministic behaviour and finally the level of description as opposed to explanation. An important basis for these typologies is the increasing openness of the systems. Isolated systems are an ideal type, ones where there is no movement of matter or energy across system boundaries. Closed systems permit the flow of energy across boundaries, but not matter.

Open systems permit the flow of both energy and matter across their boundaries. These distinctions are important as they begin to define the expected behaviour of systems. The definitions derive from physics, where an isolated system will tend towards an equalization of the distribution of energy within it and hence eventually exhibit maximum entropy and disorder or randomness in the organization of its components. Systems that have their boundaries open to flows – particularly if these are of both energy and matter – are able to stave off this 'entropy death' as they retain their organizational structure. Open systems are viewed as able to retain both entities and relationships by maintaining gradients of energy levels between different system components.

Flows from high to low energy are maintained and so the entities and relationships, the network that produces the system, are maintained. Energy and matter are derived from beyond the boundaries of the system to maintain that system. Although the overall entropy within the universe may be heading towards a maximum level, the smaller system being studied is able to reverse this trend within it by importing energy and matter to maintain its order. Entropy is kept at bay by continually exporting disorder and 'borrowing' energy to keep order.

Each of the above classifications implies an increasingly structurally complex view of the world using systems. Within each classification scheme, the simplest level is viewed as description and definition. From this level the structure of the system is identified (the morphology of the system) to which flows of matter and energy are added (the cascading system). Resultant from these two levels is the process–response system, where energy and matter flow through a set of entities arranged in a specific manner. The flows interact with the entities to produce change or stability in these entities and their relationships to each other.

What complexity means, however, is rarely made explicit. There seems to be an implicit acceptance that the lowest level in the hierarchy represents 'mere' description. At this level it seems to be assumed that there is no real explanation. Setting up the system, defining the entities and relationships to be modelled, seems to be viewed as a relatively simple and uncomplex task. This is the level upon which all other levels depend and which forms the basis for 'real' explanation.

Complexity seems to be associated with increasing refinement in the specification of entities and relationships. Likewise, a more complex system is one where the entities and relationships are dynamic in the sense of being specified and identified and measured as they change. Full understanding of the system implies that explanation can be generalized and applied to all systems with the same entities and relationships. The problem that full specification would define the uniqueness of the system, and so the uniqueness of its explanation, is not considered as an issue.

One of the most important impacts of systems thinking in physical geography has been the framework it provides for directing and organizing thinking about reality. Thinking is not directed at an individual entity but at the relationship of that entity to other entities and the context of the entity within a system. Not only is thinking directed, but it can also be represented in a formal manner. Representing reality by symbols that can be applied to different systems provides a strong unifying bond between different parts of the subject. This gives the impression of a uniformity of approach and purpose that, although it may be illusory, provides a myth of commonality that is lacking in human geography. It also could be interpreted as a singularity of approach and therefore misidentified as a lack of conceptual and philosophical breadth and depth by human geographers. Symbolic representation of diverse systems also means that they can be modelled, conceptually and mathematically, and so general relationships established and tested between systems. In this manner trends and patterns identified in one part of reality can quickly be transferred and assessed in another part of reality. Systems analysis, from this viewpoint, has helped to identify the holistic and universal nature of physical reality more rapidly and accurately than any technique before it.

Systems thinking and application in physical geography has not been without problems, however. Some of the most important issues relate to the points raised in previous chapters concerning reality and how it is viewed through different philosophies. Central to the problems of systems thinking is the potential to believe that the system is reality. A system is only a model of reality; it is a simplified representation of what the researcher believes to be real and important for the operation of the particular small area of reality they are concerned with. Any system, its components, its relationships and behaviour can only be identified and understood in the context of the theory or theories that have informed its construction. Identifying and maintaining the link between theory and the system constructed should be at the forefront of the mind of any researcher. Without this link the rationale for system construction, and the expectations for system behaviour, is unknowable. Assuming that the system is reality could result in the same system framework being applied inappropriately to different parts of reality.

Systems and Change

Within the application of systems thinking in physical geography an important property that has been developed is the ability to identify and predict certain types of behaviour. In particular the ability to identify when and why systems change or remain stable has become a focus of study. Within systems analysis the types of changes expected are related to how the system and its components change as the inputs change. As inputs, usually considered as discrete events, change in their magnitude or frequency, or both, it is expected that the output from the system or even the internal organization of the system will alter. The type of change will depend upon how the system is organized and the relationships between variables. Importantly, the presence of feedbacks within the system will determine system behaviour. Negative feedbacks will tend to dampen the impact or tendency to initiate change in system behaviour.

Positive feedbacks will tend to enhance the changes caused by any input in system behaviour. Implicit within most assessments of system behaviour is the assumption that the system will behave in a specific and predictable manner. This usually means that the system will exhibit some form of equilibrium behaviour. Equilibrium itself is a complicated and increasingly contentious concept within physical geography, and full discussion of it is left until the next chapter. Chorley and Kennedy (1971) view equilibrium as the maintaining of some kind of balance within a system, whether that be of the relationships amongst the variables, the level of the output or some steadily moving set of conditions. This highlights the idea of equilibrium equating to balance and being the expected or 'normal' behaviour of the system.

The idea and significance of change in the physical environment did not start with systems analysis, but the framework provided by systems thinking has been useful for formalizing the different ideas about change. To clarify how systems analysis has helped to codify change and stability it is useful to review how time has been seen within physical geography and then how change has been incorporated into physical systems. Change (and stability) can occur in both time and space, although it is temporal change that has been the focus of most study. Discussion of temporal change has been seen as a key ingredient in defining contemporary physical geography. Schumm and Lichty's (1965) paper on 'Time, Space and Causality in Geomorphology' is still viewed as a classic presentation of different scales of temporal change, even if the theory is untestable.

They identified three types of time: cyclical, graded and steady-state. At each temporal scale different variables will be important for the operation of the physical system. As the scale changes what were previously important variables effectively become constant, unchanging variables, the context of the system. Although Schumm and Lichty were aiming their critical review at the Davisian model of landscape development, their use of concepts from other parts of science such as equilibrium in developing their argument provided a theoretical basis for changing the emphasis of what geomorphologists did. They highlighted the need to focus on the appropriate scale of study and the variables appropriate to that scale.

Davis was not disproved, merely regarded as irrelevant for more process-orientated studies which focused on reality at a different scale from Davisian landscape evolution. In this respect, although

Schumm and Lichty did not use system terminology, their study highlights the importance of appropriate boundary and entity definition for studying reality. Likewise, the use of diagrams such as the one above provided a set of expected behaviours for the scale of reality being studied. They had started to define what type of changes should be expected in particular variables under certain conditions. They had started to define systems behaviour in geomorphology.

Wolman and Miller (1960) looked at how the magnitude and frequency of events could be used to understand change and stability within the physical environment. Although they limited their definition and exemplification to fluvial systems in temperate environments, their form of analysis did provide an insight into change. They identified the magnitude of an event by how much 'work' it did in a catchment. Work was defined by amount of sediment moved by events of given frequency. Importantly, the sediment moved had to be measured as output from the catchment to count as having been 'worked on' by an event. A 'normal' event was viewed as bankfull discharge. This definition limited work to a relationship with sediment movement out of the catchment. Movement within the catchment, or work done in weathering material for transportation, was not included, for example. With this definition of work it is clear that large events do most work – they move most material.

Wolman and Miller also noted that events of a given magnitude occur with differing frequencies. Large events are relatively rare, small events are common. Combining the two trends for events, they produced the magnitude/ frequency curve. From this curve they identified that medium-size events are doing most of the work within a catchment. They linked these medium-size events to bankfull discharge within the catchment and identified forms such as river banks as being adjusted to these size events. They were, in other words, defining expected behaviour for the catchment system in terms of river bank morphology and 'normal' events. The relation between form and event was, however, defined only in terms of the absolute magnitude and frequency of the event. The form and its reaction and interaction to an event is assumed to be constant, the system is devoid of feedbacks. They did, however, draw a distinction between events that may be formative, that may produce landforms in the landscape, and those that did the 'normal' work. In other words they recognized the potential for qualitatively different types of events in the landscape.

Wolman and Gerson (1978) developed a more complicated version of the relationship. In this paper they define event effectiveness – not in absolute terms but in relation to the ability of the landscape to 'restore' itself. This assumed a tendency for the landscape as a whole to move towards a preferred characteristic or equilibrium state once a disrupting event is removed. The event is still viewed as a discontinuity to the system, as an interruption to the 'proper', normal behaviour of the system. Once the event disappears, the system, in their case the landscape, can once again return to its normal modes of operation. Events are not seen as important parts of how a system operates in these papers, instead they are seen as disruptions to be endured.

Event effectiveness did not solely depend upon the property of magnitude. The context of occurrence was vital. If two large events occur one after the other, the second event would have no sediment to remove from the catchment as the first event would have already removed it. Event sequences became important, so the event in the context of other events had to be considered. In addition, an event of a given magnitude would have a differential impact depending on the power of the restorative forces in a catchment. Although Wolman and Gerson tended to limit these forces to vegetative regrowth, the concept could be applied more widely. This new definition highlights that the context within which an event occurs is important for the relationship between event magnitude and frequency and landform change. In other words, the behaviour of the system depends upon the system as much as it does upon the input into the system.

The presence of strong negative feedbacks can dampen the changes initiated by an event. Weak negative feedbacks reduce the ability of a system to 'repair' the damage of an event. It might even mean that the system could not return to its previous 'normal' mode of behaviour. Brunsden and Thornes (1979) added a further dimension to expected system behaviour in their version of landscape sensitivity. They defined landscape stability as a function of the temporal and spatial distributions of the resisting and disturbing forces within a landscape. This concept can be expressed by the ratio of the magnitude of barriers to change and the magnitude of the disturbing forces. The sensitivity of landforms to internally and externally generated changes can be expressed through the transient-form ratio:

$$TF = \text{mean relaxation time} / \text{mean recurrence time of events}$$

This definition highlights the significance of both temporal and spatial changes in a system due to an event. Events are no longer simple single inputs, their nature can change and in changing their nature the behaviour that they induce changes. Likewise, systems are no longer viewed as simply responding to an event, the event is instead mediated through the system and the mediation itself becomes a form of creeping change within the system. Events still have impacts, but the impact is more dispersed and complicated than previously envisaged. System behaviour still adheres to the idea of equilibrium, but the nature of change depends more upon the interaction between system state and input rather than just input. In Brunsden and Thornes different time-scales for individual landforms become important. Time can be divided into the time taken to recover from a disruption (relaxation time) and the time period during which the 'normal' or characteristic form is present.

The characteristic form is the form the landform takes when it is able to absorb disruptive events. Relaxation time is a time period when a landform is adjusting to a disruption and when it exhibits non-normal forms. These transient forms are not in equilibrium with the processes forming the landform; in other words, they are not characteristic forms. Landforms will have different stabilities and react to the same disruptions in different ways depending on if they are adjusting or adjusted to disruptions. Figure 6.4 illustrates this idea by viewing the landform as being entrenched to differing degrees in its current state. Different parts of the landscape will have different degrees of entrenchment and so different sensitivities to the same event.

Barrier to Change

In this way Brunsden and Thornes provide a very complex and spatially and temporally differentiated view of landscape and landform behaviour, but retaining at heart the idea of a tendency to some equilibrium condition. From the above the significant aspect for the behaviour of the system is the relationship between the resisting and the disrupting forces. These general concepts can only be made operational by the definition of the researchers. There are no hard and fast rules as to how to define each of these properties, but their conceptualization is vital for understanding how systems behave. Brunsden and Thornes's analysis requires that entities in the system be defined in relation to an input or event as either resisting or enabling its mediation as a disruption to the 'normal' operation or behaviour of the system. In this manner, the definition of normality or expectations of system behaviour become essential to defining what to study and how to understand changes in system behaviour.

Understanding the system is inherent within the manner in which the system is constructed. Schumm (1979), in the same volume as Brunsden and Thornes, further developed the concept of thresholds in a system and the potentially complex behaviour that even simple thresholds could produce. Thresholds make points in the system where the behaviour changes from one mode of operation to another. In other words, the system stops following one expected course of behaviour and, after some brief odd behaviour, settles down into a different but expected (i.e. predictable) course of behaviour. Thresholds can be extrinsic, a system responding to an external influence. Thresholds can also be intrinsic, with the switch in behaviour occurring without any change in the value of the variables external to the system. Adjustment of a landslide to the continual process of weathering of its regolith could be viewed as an internal threshold. At some point the continual operation of weathering agents could weaken the strength of the regolith sufficiently to induce a failure during a rainfall event that previously would have passed without incident. Schumm identifies a geomorphic threshold as one that is inherent within the system, one that develops through changes in the morphology of the landform itself.

A system does not necessarily respond rapidly to a threshold being crossed. Schumm uses the example of an experimental channel where there has been a single change in base level. Incision at the river mouth progresses upstream, rejuvenating tributaries and removing previously deposited sediment. As this erosive wave progresses upstream, the sediment load increases in the main channel with consequent deposition and aggradation in the previously incising channel. Eventually, the tributaries become adjusted to the change in base level and the sediment supply to the main channel dries up.

This initiates a new phase of channel incision. From a single event, the change in base level, a series of incised channels and terraces have been formed and different parts of the catchment have responded at different rates and in different ways to the same stimulus. This experiment, plus the example of Douglas Creek that Schumm uses, implies that even a simple system can have a complex response to events. The precise sequence of changes will depend on the context of the system and its thresholds. Importantly, there remains the problem of identifying sensitive parts of a system and the thresholds before changes occur.

Systems and Landscape Sensitivity

Landscape sensitivity as expounded by Brunsden and Thornes (1979) is an important means of understanding systems and change. In explaining the idea, and in developing its application (Brunsden, 1990, 2001), use is made of major concepts in systems thinking. The application of these concepts provides a systems framework that could be applied to most phenomena in the physical environment. From a Popperian perspective this could call into question the basis of landscape sensitivity as a theory, casting it instead as a myth. In Brunsden and Thornes (1979), four fundamental propositions concerning landscape development are put forward as the basis for constructing the idea of landscape sensitivity. These propositions could be viewed as basic postulates of the theory, or as untestable truisms in the sense put forward by Haines-Young and Petch (1986).

Brunsden and Thornes (1979) view landforms as either being in equilibrium with environmental conditions (characteristic forms) or as forms moving towards that state (transient). The landscape is continually in a state of change as external inputs force it to adjust towards its characteristic state or as internal thresholds are crossed. The distribution of disruptions, and therefore of landforms in different states of transience or stability, is both spatially and temporally complex. The result is a complex assemblage of landforms in varying stages along a sequence of adjustment. The susceptibility of any specific part of a landscape can be expressed by the relationship between the forces resisting change and those forcing it. In other words, there are spatial and temporal variations in the balance of the landscape and its components. Sensitivity expresses how close the landscape and its components are to the edge of this balance.

Proposition 1: For any given set of environmental conditions, through the operation of a constant set of processes, there will be a tendency over time to produce a set of characteristic forms.

Proposition 2: Geomorphic systems are continually subject to perturbations which may arise from changes in the environmental conditions of the system or from structural instabilities within. These may or may not lead to a marked unsteadiness or transient behaviour of the system over a period of 102–105 years.

Proposition 3: The response to perturbing displacement away from equilibrium is likely to be temporally and spatially complex and may lead to a considerable diversity of landforms.

Proposition 4: Landscape stability is a function of the temporal and spatial distributions of resisting and disturbing forces and may be described by the landscape change safety factor, here considered to be the ratio of the magnitude of barriers to change to the magnitude of the disturbing forces.

Brunsden (2001) emphasizes that landscape sensitivity permits an assessment of the likelihood that a given change in the controls of a system will produce a sensible, recognizable, sustained but complex response. The image of the landscape system is one that is in a state of possible change. Whether change occurs or not depends upon the characteristics of the system, but always in relation to the disruptive forces affecting it. The landscape system is balanced between its propensity for change and its absorption of the disruptions within its existing structure. Brunsden (1993, 2001) outlines several sources of resistance to or absorption of disruptions.

These usually reflect the structure or morphology of the system under investigation. System response will also depend on the sequence of events experienced. Different sequences will activate different pathways in the system and so create different responses. As the system responds to events it alters its own characteristics, and so what Brunsden calls 'time-dependent preparation processes' can alter the sensitivity of the system. Weathering, for example, can weaken the bonds within a rock mass and increase its susceptibility to failure by future disruptive, triggering events. Brunsden refers to Crozier et al.'s (1990) model of slope-ripening by increasing regolith thickness through weathering as an example of a preparatory process. Operating in the opposite direction, changes in the system can increase its resistance to disruptions. Church et al. (1988), for example, suggested that the coarsening of bed surface textures occurs in tandem with a rearrangement of the coarse clasts into increasingly stable geometric arrangements.

System Properties Description

Strength resistance Barrier to change imparted by the properties and dispositions of the materials out of which the system is made. Morphological resistance Variable distribution of potential energy across the system. Scale dependent depending upon the processes considered. Structural resistance Design of the system – components, topology, links, thresholds and controls. Subdivisions are location resistance (location of system elements relative to that of the processes capable of generating change), transmission resistance (ability of system to transmit impulse of change). Filter resistance System control and removal of energy from landscape – shock absorbers of the system. System state resistance Ability of system to resist change because of its history. Past history will have configured system pathways in a specific, unique manner, so no two systems will respond the same to the same input.

Development of landscape sensitivity has resulted in a vast number of related ideas that emerge as logical extensions of the initial propositions. The couching of each in systems terminology provides a means of incorporating these into systems thinking. The range of ideas, however, is almost too inclusive. Little in terms of change or stability does not come under the descriptive preview of one of these ideas. Testing the presence of a specific type of change is difficult. In the case of preparatory process, they can only be identified after the change; before the change they are only one of many possible processes of change. Similarly, resistances are defined by the nature of the system, an entity identified and populated with elements by the observer. System resistance is built by the knowledge and experience of the observer, but it is unknowable if there is a basis or even correspondence for this in objective reality.

SUMMARY

Systems analysis has been a dominant method of thinking about the physical environment since at least the late 1960s. Systems analysis provides a set of formal and standard terms for translating the physical environment, a common framework for analysing the whole of the physical environment. Systems are still, however, a simplification of reality, not reality as it really is. Despite some discussion of the meaning of the term, systems thinking has been applied extensively. Systems thinking has, however, constrained the type of behaviour researchers expect from the physical environment. Specifically, features such as positive and negative feedbacks and equilibrium are assumed to be present and to be 'explanations' for behaviour. Likewise, there has been a tendency to use and apply system terms, such as 'robustness', 'sensitivity' and 'relaxation' to the physical environment. These terms have very flexible definitions, depending upon the context in which they are applied. This makes the testing and falsifying of theories expressible in these terms problematic as many translations of these terms are possible.

7

CHANGE AND COMPLEXITY

EQUILIBRIUM: AN EX-CONCEPT

Understanding change in the physical environment has meant understanding and applying the concept of equilibrium. Equilibrium and its validity is the focus of a continuing debate amongst geomorphologists. Despite the disquiet with the notion, it still has a powerful hold over how physical geographers think about reality. Even where alternative concepts are offered, they often refer to equilibrium to illustrate their difference from it. Equilibrium acts as a reference point for debate, even for its opponents. Given the important position of equilibrium, it is vital to understand what it is and how it has been applied and so affected how physical geographers think.

Equilibrium is a concept borrowed, as usual with physical geography, from the 'hard' sciences of physics and chemistry. As with most appropriated concepts, its application in physical geography has not adhered to the stringent definitions applied in these subjects (assuming they ever were applied stringently in the first place). For example, within statistical mechanics equilibrium is viewed as the most probable macrostate of a system composed of potentially different microstates (Welford and Thorn, 1996, p. 670). Howard (1988) makes the point that the definition of equilibrium relies on a great deal of subjectivity in the sense that the researcher often defines the input and output variables in a search for equilibrium. Similarly, the variables selected for measurement must be capable of changing over the time and space scales of the study. Likewise, Howard (1988) also emphasizes that just using the term 'equilibrium' does not make a statement about cause and effect within a system. Instead it describes a presumed state without necessarily explaining that state.

Renwick (1992) introduces the term 'disequilibrium' to describe landforms that tend towards equilibrium but have not had sufficient time to reach this condition. This is contrasted with non-equilibrium landforms that can undergo rapid and substantial changes in form such that there is no average condition to which they seem to tend over time. In this manner Renwick distinguishes two types of opposites to equilibrium rather than a single opposite.

Equilibrium implies some sort of balance as well as the maintenance of that balance. In other words, equilibrium implies both a condition for a system and a behaviour for the system in order to maintain that condition. Within physical geography, this idea of balance has been applied at a variety of scales. It has been used to refer to the overall state of the system (unchanging in terms of inputs and outputs), to describe the relationship between inputs and outputs (they are the same), as well as to describe the status of individual variables. Some researchers have claimed that 'equilibrium' is a term that should only be applied to the last of these states. More contentious has been the extension of equilibrium to apply to systems that are changing in their states.

Equilibrium is also used to describe trajectories of system behaviour. Within a systems framework the interaction of variables will produce changes in both the variables and the system itself. If measured outputs from the system are taken to represent its behaviour then these outputs for any given time period can either remain the same, increase or decrease. Plotting these changes over time indicates the trajectory of system behaviour. A similar set of changes can occur at the level of individual variables or between variables. Schumm and Lichty (1965) make use of this expanded definition of equilibrium in their analysis of time and causality in physical geography. This view of equilibrium has spawned a range of different types of equilibrium such as meta-stable, dynamic, and quasi-dynamic. Interestingly, there is even non-equilibrium, using equilibrium behaviour to define alternative behaviour by its absence. Each of these definitions is possible to apply, because equilibrium is viewed as the typical behaviour of a system, but without any detailed explanation of why this is so. Using a systems framework implies that it may be possible to identify feedbacks and thresholds within a system and, importantly, these will influence how the system behaves.

Equilibrium relies on the presence of negative feedback loops within the system. This means that researchers begin to define an expected behavioural pattern for systems. Inputs may vary, but the system will respond by tending to dampen the impact of these changes and returning to its previous condition. The system will have a 'normal' behaviour and tend to want to return to it. This expectation raises issues of teleology in system functioning, i.e. that the system behaves as if it had a predefined function. These types of behaviour are generally explained away in terms of system tendencies rather than imputing functional behaviour to a system. The expectation of an identifiable 'normal' state for any physical system does imply that change is unusual in reality. Researchers will therefore tend to look for evidence of change being reduced rather than change being amplified. The latter suggests positive feedback which will create unstable structures within systems.

Stability of entities and what appear to be coherent systems to the researcher imply that reality must be dominated by negative feedbacks. By 'disturbance' researchers have tended to mean events, or rather inputs into the system. A range of different types of inputs have been identified (e.g. ramped, pulsed, continuous) which can all influence system behaviour in different ways and over different time-scales. If system equilibrium were defined as a fixed relationship between input and output or between variables and their states then the concept would have limited application in reality. Inputs vary both as discrete units and as continuous values, and as a result so do outputs. There are fluctuations in both, and hence in the variable states that produce them. Ironing out these variations has been the basis for identifying equilibrium as an average state of a system. Once an average is involved the idea of equilibrium as a user-defined state begins to become clearer.

An average implies the selection of a time-scale (and more rarely a spatial scale) over which values for the system state can be averaged. Likewise, it also implies a series of measurable system states at a temporal resolution (or spatial resolution) shorter than the averaging time period. In other words, as noted in previous chapters, the researcher makes the world, the reality, of the equilibrium via their measurement systems and theoretical constructs. Equilibrium becomes identifiable because the researcher is trying to find it.

Defining equilibrium by an average system state provides the basis for defining steady-state equilibrium. It also, however, permits the extension of the concept to situations where the average state is not constant. Although inputs and outputs may change, the structure of the system may remain constant. The morphology of the system may remain constant even if the flows through it increase or decrease. In this sense, the system can be said to be stable. Researchers have tried to capture this stability of morphology as opposed to constancy of input and output values by the development of concepts such as dynamic equilibrium.

Dynamic equilibrium refers to the progressive change of a system around an average fluctuating state. This type of system behaviour appears to show the system changing, and indeed it is. The system structure, how flows are organized and how disturbances are nullified, remains constant. What alters is the value of variable states within the system. Within a slope system undergoing slope decline, for example, the manner in which material is moved across the slope surface, the transfer of energy from rainfall input to movement of soil material, does not alter. What alters is the rate of this movement. Flows alter because state variables alter; in the case of this example, slope angle slowly declines and so other connected variables such as slope length and height alter. The state of system variables is therefore undergoing constant and predictable change.

In no manner could this be described as a steady-state equilibrium, even in terms of an average state. This does not, however, capture the point that the operation of the system remains the same. Dynamic equilibrium allows this point to be conceptualized. It highlights that change can occur in the system, but that stability is also preserved in how the system functions. Once again the idea of a system tending towards a normal mode of operation, and this stability being the normal state of the physical environment, is central to systems thinking. Preservation of system morphology is also a theme within the development of definitions of equilibrium such as 'meta-stable equilibrium' and 'quasi-equilibrium'. On occasions systems change their behaviour dramatically. Schumm's (1979) identification of internal thresholds within systems implies that progressive change may result in the crossing of an irreversible boundary. A point of no return is reached where the structure of the system breaks down or begins to operate in a different manner.

At this point the previous expected types of behaviour disappear. They are replaced by a new 'mode' of behaviour: a new normal condition is established for the system. The flip between states is the meta-stable part. The system approaches a threshold in either its overall behaviour or for individual variables. Crossing this threshold changes the relationships between variables and a 'new' system state is born. The identification of this change is up to the researcher. It may involve the redefinition of the system structure to include new variables or the redefinition of how variables interact or what they are. The important point is that there is a user-defined restructuring of the system. Despite this restructuring, there is still an expectation that the 'new' system will continue to attempt to establish and maintain an equilibrium, that a 'normal' set of behaviours will emerge and function. At this point it may be possible to begin to define domains of expected behaviour. The possible trajectories for a system can be mapped out in terms of this behaviour space. This means that researchers will expect system behaviour to focus about these 'norms'.

Of equal significance, researchers will predict what changes will result in which equilibria being reached and how long a system might occupy a specific equilibrium. In other words the physical environment becomes a predictable entity. Brunsden and Thornes (1979) use 'characteristic' forms to represent landforms in equilibrium with their environmental conditions and 'transient' forms as landforms tending towards characteristic forms and so not in equilibrium with their environment; both are examples of the pervasive grip of equilibrium. Characteristic forms are viewed as the 'normal' forms, whilst transient forms are ephemeral, unexpected and not 'normal'. This underlying set of expectations is further enhanced by the use of both terms in defining landscape stability – ratio of characteristic as opposed to transient forms. In other words, the ratio of normality to oddities in the landscape. The greater the oddities, the less stable the landscape. The expansion of the range of definitions of equilibrium means that the opposite of what equilibrium is becomes harder and harder to define.

The flexibility, fuzziness and adaptability of definitions of equilibrium are a key strength of the concept and also its key weakness. Some have claimed, and not without cause, that equilibrium can mean anything and so is of little use as a workable and explanatory concept for understanding the physical environment. As the sphere of system behaviour encompassed by the term 'equilibrium' increases, it becomes increasingly difficult to determine any sort of behaviour that cannot be redefined as some sort of equilibrium. Meyer's (1992) definition of non-equilibrium forms still uses the standard system behaviour diagrams to illustrate how unlike equilibrium his non-equilibrium forms are. Such problems of trying to define an alternative to equilibrium illustrate how common and constraining the thinking developed by the concept has become in physical geography. The emergence of chaos theory and complexity in the last twenty or so years has suggested that there may be other ways of thinking about the physical environment and how it changes, but the use of these concepts has shown how ingrained equilibrium is within physical geography.

Chaos and Complexity: More of the Same

Chaos and complexity are not the same thing. Although Manson (2001) recognizes that there are different types of complexity, each with different and often contradictory assumptions, he still attempts to provide a threefold classification of complexity. Algorithmic complexity refers to complexity as the simplest computational algorithm that can describe and so reproduce the behaviour of a system (Manson, 2001, p. 406). Deterministic complexity is related to chaos in Manson's typology. This type of complexity displays sensitivity of system trajectories to the initial system conditions. It also assumes the possibility of attractors, to which a system trajectory tends, in amongst the seemingly chaotic behaviour. It is this form of complexity that Manson feels appeals most to post-modernist human geographers. Aggregate complexity refers to the holistic behaviour resulting from the interaction of system components. These interactions produce emergent behaviours, behaviours that cannot be predicted from the individual components alone. The basis for this emergent behaviour lies in the interaction of system components.

Reitsma (2003) provides a critique of Manson (2001). She suggests that Manson's typology is one that identifies different measures or definitions of complexity rather than different types. She also carefully draws a distinction between complicated and complex. A complicated system is one where a complete and accurate description of a system can be given in terms of its individual component parts. Although such a description may contain a lot of information, a lot of data, the operation of the system can be predicted from its parts. Indeed, in the examples Reitsma presents of a computer and VCR, the predictability of the machinery is the whole point of their construction. In contrast, complexity refers to a system where the description of components does not provide enough information to predict the behaviour of the system. She suggests that often complexity is assigned to complicated systems. In a reductionist view of the world complicated systems are likely to be the norm, as parts predict the whole, whilst in a world of complexity emergence reigns and the whole is unique relative to the parts.

Reitsma also states that chaos and complexity are not presented as clear and distinct theories within Manson's discussion. Reitsma accepts that there is no common framework or set of definitions for distinguishing chaos and complexity, but the two need to be clarified as different. Chaos theory is concerned with the operation of simple, deterministic, non-linear, dynamical and, importantly, closed systems. It is these systems that are sensitive to initial conditions, and which can produce seemingly chaotic behaviour under the action of slight perturbations. Complexity theory is concerned with complex, non-linear and, in contrast, open systems. Complex systems, rather than 'degenerating' into chaotic behaviour, respond to a perturbation by organizing their components into emergent forms that cannot be predicted from the system components themselves. This is system self-organization.

Algorithmic complexity Simplest computational algorithm that can reproduce system behaviour. Complexity lies in difficulty of describing system characteristics mathematically. Deterministic complexity Interaction of variables produces systems that can be prone to sudden discontinuities. Sensitivity to initial conditions and bifurcation are key characteristics. Aggregate complexity Individual components of a system operate to produce complex behaviour. Deterministic complexity Based on information theory. Algorithmic content of a string of bits. Complexity equated with randomness. Statistical complexity Measure degree of structure present. Randomness equates to maximum complexity. Phase transition Maximal complexity is mid-point between order and chaos.

Chaos derivatives Precise definition through indices such as Lyapunov components – system's sensitivity to initial conditions. Connectivity Complexity is measure of degree of connectivity in system. Greater connectivity equates to greater complexity. Oddly, this may mean greater system stability. System variability Increase in system variability reflects an increase in complexity. Relative and subjective complexity Complexity arises because of human perception and so only exists relative to the observer. Relationships and change within the physical environment. Most introductions to these concepts define them as contrasts to the Newtonian view of the world. Although this caricature is shorthand for a particular linear modelling view of the environment, it is also a straw-man that is probably an inaccurate description of how most researchers perceive the physical environment anyway.

Nonetheless, the image of physical reality painted by the linear view does sound familiar. Variables interact in a regular manner and we can determine patterns to this regularity that we can model so as to predict future behaviour. An important part of this transferability of prediction is the idea that system states that are close together will evolve or change along similar, if not identical, trajectories. Chaos theory holds that systems may be extremely sensitive to their initial condition and that change may be non-linear rather than linear. Both these conditions imply that systems will behave in an unpredictable, if not chaotic, manner.

Even if the initial state of a system can be measured, the accuracy of that measurement may be insufficient to preclude such behaviour. In this sense chaos theory goes against the grain of scientific thinking since Newton. It begins to suggest that some 'laws' of nature may be unknowable to us because they produce irregular behaviour in our user-defined physical systems. Phillips (2003) provides a set of common types of complex non-linear dynamics that provide a wide coverage of different types of system behaviour. Behaviour of most systems observed in the field could be interpreted within the framework of chaos and complexity, or, as Phillips prefers, non-linear dynamical systems.

Phillips (2003) also identifies that non-linearity has a variety of sources within the environment. His definition of why the sources are non-linear, as well as the examples used, highlights the continuing view that geomorphologists have been studying non-linear relationships all along, but have not recognized these as such. Importantly, within Phillips's scheme is the point that the presence of non-linearity does not necessarily mean that chaotic or complex system behaviour will occur as of necessity.

Complexity is not relevant to every geomorphic problem even if the potential for it to occur is present. The sources of non-linearity are varied and have often been explored and explained by other means by geomorphologists. This does not negate the potential importance of nonlinearity for providing other types of explanations for complex behaviour, but it means that non-linearity need not be the obvious or immediately clear source of complex behaviour. Non-linearity is only one tool in the explanatory kit of physical geographers. Within physical geography chaos and complexity have been mentioned, but still not widely embraced. Where they have been seriously assessed it is usually in relation to existing concepts such as equilibrium. Phillips (1999), for example, provides an outline of some of the definitions used within complex non-linear dynamics. The two concepts, equilibrium and chaos and complexity, seem to produce similar definitions.

Phillips's analysis seems to suggest that physical geographers had been using complexity all along, but just had not realized it. This view suggests that chaos and complexity theory is already operating within physical geography, so there is little that physical geographers need to do to incorporate it fully. It also implies that the types of explanations offered by chaos and complexity theory are already familiar to physical geographers, and so little needs to be done to adapt physical geography to these new ideas. Alternatively, it could suggest that there is little point in physical geographers adopting the new concepts as they add very little to the explanatory frameworks they already have. There may be something to this point, but it may also be fair to say that the explanatory framework suggested by these two theories is one that a lot of physical geographers would feel uncomfortable with.

Just stating that a system is chaotic and so unpredictable does not sound much like an explanation. It could be that this negation of explanation, as some may interpret it, is part of the resistance to accepting chaos and complexity theories. However, it could be argued that equilibrium is little better. Assuming a tendency towards a 'normal' state, even when dressed up with terms such as 'negative feedback loops', is still a constraint on explanation. Acceptance of inherent uncertainty as a valid form of explanation does seem to be almost anti-explanation.

Likewise, the acceptance of chaos and complexity implies a change in what is defined as explanation in physical geography. Identifying chaos implies inherent instability in the operation of physical reality. It implies that a simple linking of one event to another, of linking a cause to an effect, may no longer be possible. Within systems analysis and the concept of equilibrium, a causal chain could be set up between events and outputs. The chain may be long or short, resistant to disruptions or sensitive to change.

Within chaos, events now have multiple effects, singular patterns of cause and effect are potentially lost, and the context of the causes becomes important. A system close to chaos will produce different outcomes to one far from chaos. Although an argument could be made that equilibrium explanations have a similar type of change when the system is near a threshold, chaos regards these multiple relationships as inherent within the system – they are normal, they are not reflections of behaviour at the extreme of the system. In other words, chaos and complexity begin to redefine what is normal for a system. Identifying variables, their states and relations no longer guarantees that certain modes of behaviour will occur or can be predicted. As well as unpredictable behaviour, concepts from complexity theory have also been applied to try to understand how order, in particular self-organization, has developed (e.g. Dunne et al., 2002 and Williams et al., 2002 in relation to food webs). Early work by Kaufmann on the self-organization of molecular systems indicated that at a critical threshold connecting entities within a system resulted in the emergence of 'organization'. Initially, the system of 'buttons' is relatively weakly connected.

Adding more 'threads', a point is reached, as the ratio of threads to buttons reaches 0.5, where suddenly the whole network of buttons becomes interconnected. The sudden jump in connectivity was reflected, Kaufmann believed, in molecular systems that were autocatalytic, that had self-reinforcing feedback loops for individual molecules. These molecular systems became very stable and able to resist changes in the environment, and even act as relatively autonomous agents in the physical environment. Maintaining this organization or structure despite the perturbations thrown at the molecular system by the environment illustrated how robust such self-organized networks could be. Rethinking other networks of entities in the same light, questions could be asked about their ability to maintain their structure when disrupted and the relative importance of different types of nodes with different degrees of connectivity within such networks.

EMERGENCE AND HIERARCHIES: SCALE REVISITED

The development of chaos and complexity theories has, however, begun to highlight the importance of context for explanation. Additionally, the development of critical realism and pragmatism has brought alternative explanatory frameworks to the attention of physical geographers. This has stimulated consideration of some old ideas within physical geography in a new light. Significantly, the role of context in explanation has surfaced as a key issue in contemporary physical geography. The importance of reductionist forms of explanation cannot be underestimated within physical geography. However, like equilibrium, reductionism has undergone various changes in definition that make the establishment of its precise nature difficult to determine and the identification of a potentially alternative explanatory framework difficult to establish. Reductionist, refers to the tendency to reduce explanation to the level of the lowest identified entities. Within physical geography this usually involves assuming that all 'real' explanation is located at the level of physics and chemistry. At this level, explanation is viewed as generalized and applicable to all entities. Increasingly, field sciences such as physical geography have found such sentiments of universality difficult to sustain, given the questions of interest about reality in their subject. The problem seems to lie, with the fixation of physical geography to provide what are perceived to be 'scientifically valid' explanations using the lowest entities available. Explanation is reduced to what are perceived to be natural kinds, which are therefore viewed as real entities, ones that actually do exist in reality.

Increasingly, use and familiarity with philosophical positions such as critical realism and pragmatism have highlighted the errors of this reductionist view of reality and have generated, amongst other things, a redefinition of reductionism. Critical realism, in particular, has enabled a reconceptualization of what is an appropriate explanation. Critical realism recognizes that reality is differentiated and stratified, as noted in Chapter 4. Each stratum of reality is composed of entities that are composed of, but not reducible to, entities found at lower strata. Similarly, the entities in a particular stratum operate according to regularities ('laws') that are not necessarily derivable from laws at lower strata. In other words, each stratum may have unique entities which interact according to unique sets of laws as well as obeying, or rather interacting, according to laws found in lower strata. Explanation can be divided into the vertical and the horizontal. Vertical refers to explanation that focuses in on explaining one level by reference to laws at another level. This form of explanation looks for the underlying mechanisms that explain the specific power of a specific phenomenon.

For example, the reaction of acid rain with calcium carbonate can be explained by reference to valency theory. Valency theory in turn can be explained by reference to the interaction of atomic and subatomic particles. At the lowest level the interaction of these particles can be explained by quantum physics. Extending this form of analysis to its logical conclusion would be a classic reductionist approach. Everything would be reduced to explanation at the quantum level. Unfortunately, at this level all entities are reduced to quantum particles. Building up from the level of quantum mechanics to that of the entity of interest is problematic. Some practitioners prefer to locate explanation at the level of fundamental mechanics and chemical reactions. Stopping at this level, it is often argued, is more appropriate. This implies that researchers have in mind an adequate level of explanation for their studies. Adequacy and appropriateness are therefore user-defined and capable of alteration.

Horizontal explanation is concerned with linking entities at the scale of study. This contextualizes explanation. This level of explanation highlights the significance of the juxtaposition of different mechanisms, different causal structures in determining the actualization and the power of a phenomenon. The contingent sets of relationships that define why something happens in reality, rather than how, form the focus of this level of analysis. In the case of acid rain and calcium carbonate, the question is not so much why do they react, but why do they occur in the environment in that location at that point in time to be able to react? Hierarchically based explanations follow a similar differentiated view of reality and of explanation. Hierarchical explanations such as are outlined in biology and ecology (Allen and Starr, 1982; O'Neil et al., 1980; Salthe, 1985) and within geomorphology (Haigh, 1987) provide guidance as to what levels are appropriate for explanation.

The level of study is viewed as providing the entities and relations that require explanation. The level below the level of study is the level at which processes and mechanisms are located, whilst the level above the level of study provides the context. This three-level view of explanation does provide a useful starting framework, but clearly problems arise over exactly how each level is defined. What it does highlight is that each level acts to mediate the actions of the levels below it, whilst being constrained in its actions by the levels above it. The series of complex interrelationships between levels provides a much richer and more diverse set of explanations than a single and simple view of links between scales.

This form of explanatory structure has similarities to the ideas put forward in hierarchy theory, used predominantly in ecology. Within this form of explanation it is accepted that reality is divided into different, distinct layers or strata, as recognized in critical realism. The levels in the hierarchy interact and can influence each other. The rate of processes in the lower levels is much greater than the rate of processes in the upper levels of the hierarchy. Processes of photosynthesis, for example, are much faster than those of plant growth. Although the processes and entities of the upper levels are constrained by the processes operating at lower levels, they are not solely determined by, nor predictable from, the lower-level processes. Interaction, the development of 'laws' or rules of interaction specific to a particular level, also influences behaviour. Likewise, the upper levels can influence the lower levels by affecting the conditions within which the faster processes operate. Salt weathering may be rapid on a building, but it will not occur if the conditions are not appropriate for the combination of salt and moisture. Significantly, the researcher defines what is the appropriate level of study. Once this has been determined it is then only the lower and upper levels immediately adjacent to the level of study that should be used in any explanation. Moving beyond these levels is reducing or generalizing the explanation beyond what is deemed appropriate. Too general or too reductionist in mode and the explanation becomes unrelated to the specifics of the level of study.

Recent debate has focused on the possible existence of emergence as a possible means of explanation in physical geography (Harrison, 2001; Lane, 2001). 'Emergence', like most of the terms involved in explanation, is difficult to pin down accurately. Some researchers view it as referring to the tendency for new and novel entities or structures to emerge from the interaction of processes. These new entities are dependent upon the operation of processes and other entities for their existence, but their behaviour is not merely an aggregate of these other entities. Instead, the emerging entity has a coherence of form and function that defines it as a distinct and separate individual at a particular scale. The emerging form can interact with other entities and respond as a single whole. The behaviour of the entity is greater than the sum of its parts. This view of emergence sees entities as forming at distinct scales as a response to processes at other levels. Action at a particular scale is not dependent upon processes at a smaller scale but on the emergence and behaviour of emergent entities. Other researchers view emergent entities as being distinct individuals at a particular scale, but that these individuals are explainable by reference to the processes that formed them. The entities may act as individuals and even interact with other individuals at that scale, but their behaviour is an aggregation of processes and entities found at a lower scale. In this view of emergence, entities emerge, but their behaviour and existence are predictable from our knowledge of their aggregate components. Such distinctions are vital, as in the latter view of emergence explanation is placed at the level of the components, whilst in the other view explanation is placed at the level at which the entity emerges.

The above views of entities and emergence have echoes in the writings of A.N. Whitehead (e.g. Whitehead, 1929). Whitehead viewed entities as being temporary concretizations of flows of processes. Processes became solidified in the act of creating entities. The entity only remains, however, if the flows themselves remain constant. The entity could affect the flows and so enhance the solidification of its form. Any single entity, though, was connected to all other entities by virtue of the flows that formed it. This meant that a single entity was connected horizontally to other entities at a scale similar to itself. Likewise, entities were formed by processes that were governed both by internal mechanisms and by their context.

Combined, this view of emergence begins to ask interesting questions about scale and its definition in physical geography. A common assumption is that scale is concerned with an absolute set of spatial and temporal dimensions. Entities have an existence relative to this fixed frame of reference. Different entities exist at different magnitudes within this fixed frame of reference. Recent papers, such as Raper and Livingstone (1995) in relation to GIS and by Massey (1999) and Harvey (1994) in relation to geography in general, have begun to suggest that scale may not be as clearly defined or fixed as previously thought. As with Whitehead's philosophy, a more relational or even relativistic view of scale may be more appropriate to physical geography. Rather than scale being absolute within a fixed reference frame, with the entities fitting within this, scale could be thought of as being defined by the entities themselves. Scale is no longer absolute but dependent upon the entities under study. The entities themselves define the processes or flows forming them; they define the spatial and temporal dimensions of importance rather than being defined by these.

Scale is no longer an absolute quantity but one that varies with the entities being studied, and so is defined by the study itself. This is rather different from the static spatial and temporal frameworks provided by Schumm and Lichty (1965). Within their framework space and time are absolute quantities within which variables vary in significance. The fixed spatial and temporal frame provides a common and unfaltering backdrop for the action of variables. Turning the whole structure on its head, the spatial and temporal framework becomes relative, defined by the variables themselves. This provides a more dynamic view of the framework within which variables and relations operate. The scale of operation of a variable is defined by its relation to other variables, not by a fixed framework. Phillips (1988, 2001) also looks at the issue of scale within physical geography.

Following the work of Schaffer (1981), Phillips (1986, 1988) notes that it can be demonstrated mathematically that in systems where interactions operate over time-scales of greater than an order of magnitude, these can be thought of as being independent: Controls acting at any given scale can be considered to be an abstracted subset of all system components operating at all scales. If the abstracted and the omitted relationships operate over spatial scales an order of magnitude different, the relationships are independent of each other in terms of their influence on system behaviour.

This is a similar argument to that put forward for a hierarchical view of reality. In this view processes at the lower levels of strata operate at rates so fast as to have little influence on the upper strata. Likewise, the upper strata, although constrained by the lower strata, operate at their own rates and have little impact upon the state of the lower strata.

Each level has its appropriate conditions for stability that may only be influenced by the other strata in exceptional circumstances. One of these circumstances could be the presence of chaos or complexity within and between the levels. Importantly, in Phillips (2001) this recognition of scale differences is combined with a recognition that this means that there is unlikely to be a single, best representation of reality across all scales. An appropriate representation at one scale, derived from a specific method, will not form an appropriate representation at another scale. A micro-scale form may most appropriately be identified and measured by a particular methodology. This same methodology will be unable to represent the phenomenon that the entity represents across all scales. As the scale of study changes, the nature of the phenomenon changes and the means to represent it in an appropriate manner alters.

The problem becomes, as Phillips notes, one of trying to reconcile unsuccessfully different and sometimes competing methodological approaches. This is not a problem if it is accepted that reality is not capable of continuous and singular representation: Any frustration and despair arising from accepting the argument that there are fundamental limits on the ability to use single representations across scales would likely come from reductionists confronted with the realization that their methods are not always sufficient, or even worthwhile, for addressing some problems. Proponents of historical and system-oriented approaches have acknowledged that their methods are often insufficient and sometimes useless, and have generally recognized and accepted that historical or systems approaches are best suited to particular ranges of time and space and ill-suited to others.

The scale changes, and so does the means of representing phenomena. The entities that reflect the phenomena have to alter as the means by which they are identified and measured alters. Entities and scale covary. It is pointless to view scale as independent from the entities being measured; the two cannot exist independently.

Complexity and Change – Landscape Evolution and Organization

Phillips (1995) outlines how the concept of non-linear dynamical systems (NDS) can be applied to the analysis of relief evolution. The key question he is trying to answer is does landscape evolution exhibit deterministic chaos? Deterministic chaos would result in complex, irregular patterns resulting from deterministic systems regardless of the presence of other factors such as stochastic external forcing factors or confounding factors.

Phillips considers the non-linearity in landscape development to arise from a non-proportional relationship between inputs and outputs resulting from changes in stores and system thresholds. Phillips suggests that deterministic chaos would leave a signature magnifying the imprint of disturbances of any magnitude and so produce complex (or convoluted) topography.

Without deterministic chaos small disturbances would be irrelevant and quickly removed as the topography converged on a steady-state standard topography. Stated simply this means that if relief increases over time, chaotic behaviour is implied, whilst if relief decreases over time non-chaotic behaviour is implied. It is important to recognize that Phillips does not state that relief divergence (increasing difference over time) is indicative of chaotic behaviour, only that it could be. He recognizes that other factors could cause such behaviour, including stochastic forcing factors.

In a NDS of the landscape there will be n interacting components, $x_i = 1, 2, \ldots N$. Over time the behaviour of any of these individual components could be expressed as a function of other components (x) or parameters (c) in the form of an ordinary differential equation:

$$\frac{dx_1}{dt} = f_1(x_1, x_2, \ldots x_n, c_1, c_2, \ldots c_m)$$

In an ideal world, everything would be known about the system, every component and every parameter would be known and would be expressible as above. In this case how a system changed could be described by mapping these equations in n-

dimensional phase space. In reality, such complete knowledge is not possible, so a smaller subset of the n-dimensional phase space is plotted: q-dimensional phase space. Phillips introduces the concept of phase space to help in his description of system behaviour, but it is also worth noting that the recognition that a smaller subset is used in reality to map system behaviour raises questions about component and parameter selection. How representative is the subset of the larger phase space and if other components and parameters were selected would the description produced change?

Whether the landscape has a tendency towards convergence or divergence of relief can be determined by looking at the Lyapunov exponents (l). Lyapunov exponents of a system are a set of invariant geometric measures that indicate system dynamics. If system behaviour is visualized as a set of trajectories in phase space, then the Lyapunov exponents quantify whether trajectories converge or diverge and how rapidly they do this. Where the average Lyapunov exponent is positive, then the trajectories diverge; when it is negative, they converge; when it is zero there is no divergence or convergence.

The average Lyapunov exponent refers to the general behaviour of the system. A positive value only implies that chaotic behaviour will be exhibited, not necessarily that chaotic behaviour will always be exhibited. Taking the Lyapunov exponents for individual components there will be a range of values. Where positive values occur, this implies a potential for chaotic behaviour with the largest positive value indicating the potential rate of divergence.

Calculation of Lyapunov exponents is relatively straightforward if all the differential equations describing the system components and parameters are known. In reality, this is never the case. Instead, Phillips makes use of a useful property of NDS: that the dynamics of the system can be derived from a single observation. This observation integrates the mode of system behaviour and charting its change over time will reflect the behaviour of the NDS as a whole.

If the NDS is chaotic then randomly chosen pairs of initial conditions will diverge exponentially at a rate equal to the largest Lyapunov exponent. Mathematically, once two trajectories become separated it is increasingly difficult to describe the trajectories using the Lyapunov exponents, so every so often the variations in trajectories need to be rescaled, or better still renormalized, to produce trajectories that are close together. For landscape development the theoretical terms can be translated into components capable of assessment. Phillips uses two theoretical points in the landscape, *i* and *j*, with elevations h_i and h_j at two times, to and *t*. The elevation difference is given by $h_i - h_j$ and the rate of change is:

$$d_{i,j}(t_o + t) = [h_i(t_o + t) - h_j(t_o + t)] = [(h_i(t_o) + h_t) - (h_j(t) + h_j)]$$

For this representation of the landscape and its dynamics a positive Lyapunov exponent is present when:

$$d_{i,j}(t_o + t) - d_{i,j}(t) > 0$$

In other words, if the rate of change at time t is greater than the rate of change at time to then the system has a positive Lyapunov exponent and is chaotic. Using this relatively simple means of identifying chaotic and non-chaotic behaviour Phillips defines ten types of behaviour for relief, five stable, five unstable. Phillips notes that the NDS model of topographic evolution could provide a unifying framework for all existing theories about topographic evolution. All these theories could be mapped onto the ten modes of change, making the NDS model itself unfalsifiable.

Table: Relief behaviour, from Phillips (1995).

Behaviour type	Description
Stable 1	Planar surface experiencing spatially uniform rates of erosion, accretion and uplift.
Stable 2	Both points eroding, with erosion rate at the initially higher point greater than or equal to that at the lower point.
Stable 3	Both points accreting or being uplifted, with rate of increase in height at initially higher point less than or equal to that of the lower point.
Stable 4	Initially higher point is eroding and lower point is accreting or being uplifted.
Stable 5	Higher point is not changing and lower point is accreting or being uplifted.
Unstable 1	Planar surface with any variation in erosion, deposition or uplift rates.
Unstable 2	Both points eroding, with rate at the initially higher point less than that at the lower point.
Unstable 3	Both sites accreting or being uplifted, with rate at the higher point greater than at the lower point.
Unstable 4	Initially higher point is being uplifted or accreting and lower point is eroding.
Unstable 5	Higher point is not changing and lower point is eroding, or lower point is not changing and higher point is accreting or being uplifted.

Despite the un-Popperian status of the NDS theory, Phillips does think it provides some basic tenets for a model of landscape evolution. These are, first, that there are ten modes of topographic evolution differentiated by the average rates of uplift/ accretion or erosion/subsidence at initially higher and lower points. Second, that the five stable modes involve constant or declining relief over time, whilst the five unstable modes involve increasing relief over time. Third, neither stable nor unstable modes can persist over geologic time-scales or across the whole landscape. Fourth, chaotic modes involve sensitivity to initial conditions and inherent unpredictability, but within well-defined boundaries associated with relative rates of change in the landscape.

Lastly, planar surfaces are unstable. Phillips's view of NDS as a potential integrative theory for landscape development is an interesting application of ideas from chaos theory. The development of a mathematical basis to his argument permits him to identify some indicators of chaos within systems, provided that the theoretical concepts can be appropriately translated into real entities capable of measurement. It is problematic that the theory predicts almost every mode of landscape development as this makes the falsification of the theory virtually impossible.

Any pattern of behaviour is permitted by the theory and so every pattern of behaviour can be explained away. NDS does, however, provide a relatively easy to understand interpretation of landscape development using almost the simplest of systems: two points. The range of behaviour that these two points exhibit can be interpreted in terms of stability and change, but does the framework of NDS provide any more explanatory power than the traditional framework of equilibrium?

SUMMARY

Change and stability in the physical environment have been explained via systems thinking. Equilibrium has been used as a reference point for different types of behaviour – even alternatives to equilibrium are defined relative to it. 'Equilibrium' is a term borrowed from other sciences and applied in a manner that makes it very difficult to disprove. The concept can be translated in a range of ways depending on the circumstances. The use of equilibrium as an explanation has resulted in the expectation of certain types of behaviour and constrained research to look for these forms of behaviour. The development of complexity as an alternative to equilibrium has suffered from many of the same problems. Depending on the interpretation, complexity and chaos theory can be interpreted as including equilibrium as a form of expected behaviour. Terms such as 'self-organization' are capable of translation to reality in a number of ways and so are difficult to test as explanations. Similarly, identification of chaos or complexity relies on the same measurement systems used to identify equilibrium. Separating the two concepts and devising decisive tests has been problematic.

8

PHYSICAL GEOGRAPHY AND SOCIETIES

PARADIGMS AND SOCIAL NETWORKS

The idea that scientific thought and its change is strongly influenced, if not determined, by the society within which it exists is something that Kuhn developed into the concept of paradigms. Although a paradigm is not a well-defined concept (at least 54 definitions according to Masterman, 1970) it has acted as a useful beacon for studies that emphasize the importance of social context for science and, probably more contentiously, for the idea that changes in scientific ideas are illogical and totally socially constructed. Even if you do not subscribe to the very strong social or even socially determined view of change in scientific ideas, the development of science within a social vacuum cannot be sustained any longer. Science, and by implication physical geography, develops within a range of social networks which, even if they do not impact upon the logic of selecting theories or paradigms, certainly do contribute to what and how reality is studied.

Physical geographers are part of society; in fact, they are part of a number of interlocking and interweaving societies. This detail is often lacking in a discussion of both paradigms and of the social construction of reality. The assumption of a single and easy to-identify 'society' with, likewise, simple and singular 'social influences' is a fallacy that requires some correction. It may be more appropriate to think of any physical geographer (or indeed any scientist) as located within different social networks, different sets of social relationships, that can influence how they perceive, study and hence understand reality. Physical geographers are members of humanity and as such are constrained in their behaviour and thoughts. This general level is, however, not usually the sense in which the term 'society' is used in papers exploring paradigms and the social construction of knowledge. A more common use of the term is in relation to membership of either Western or capitalist society. At this scale, the scientist is often viewed as some sort of automaton playing out, more often than not in a very negative and detrimental manner, a set of capitalistic and imperialist imperatives.

This form of analysis has echoes of the simplistic analysis of imperial imperatives used in studying geography in the nineteenth century. Patterns of behaviour and study may be recognizable, but the subsuming of every physical geographer in the UK or USA as a direct and unthinking agent of capitalistic dogma, and every action and thought as being explicable in such terms, is highly reductionist and overly simplistic. Even if actions and thoughts can be interpreted as influenced by the individual being in a capitalist society, the other social networks of which the individual is a part, as well as the individual her/himself, provide the framework within which such a broad social influence is mediated. It should not be expected, therefore, that classifying individuals as from the same society at this scale confers some sort of explanatory power onto that level. Any influence is mediated and negotiated through different social networks. As with imperialist studies of geography, this can mean that different individuals can have very different ideas about what the imperatives of their societies are.

Moving down from the scale of capitalistic society, the problem of social influence upon scientific thought becomes more problematic and more interesting. An individual physical geographer is bound into social networks at different levels. The individual is a member of social networks defined by space, such as a nation or even a nation-state, as well as more regional and local communities. At the smallest scale the individual has a tight and close social network of immediate work colleagues, friends and kin. Although often only the 'professional' relations are highlighted in an analysis of social influences, the other relations do not disappear just because the researcher focuses on the professional.

Any individual will be influenced by social networks other than professional ones, if only in the amount of time they devote to these other networks. An individual will also be located within seemingly aspatial social networks such as class, economic and international networks of fellow professionals. These seem aspatial as they do not require space for their definition, although they do require a physical, and therefore spatial, manifestation for their operation. As the scale at which social networks are considered becomes increasingly refined, the detail required to understand the networks increases and the potential range of influences increases. The individual will respond and act within these different networks in different and not necessarily consistent ways. This makes the unravelling of a model of general social influences extremely difficult. The uniqueness of the individual and their interactions with their individually constructed social networks make such generalizations difficult and often irrelevant to understanding how ideas develop and change.

The above is not an argument for neglecting the social influences on the development of ideas in physical geography. Instead it is a plea for a more subtle and thoughtful approach to what the nature of this influence might be. The individual physical geographer should be placed in their complete and complex social networks rather than caricatured as unthinking and unfeeling passive agents of social trends. It is as responsive and emotive individuals that physical geographers interact within their different social networks, and indeed with different physical environments, and from this interaction that their beliefs and opinions concerning concepts in physical geography develop. Despite the above comments, some general models of the influence of social networks on physical geographers and the development and selection of ideas have been developed. Stoddart (1981) provided a very general model of the hierarchical nature of the social networks within which physical geographers are embedded.

The important aspect of the model is how the individual mediates the social influences. Not all social influences are the same, however. Some social practices may not be viewed as such as they have become ingrained within the social networks as expected norms or rules of behaviour. This is equivalent to the 'disciplinary matrix' identified by Kuhn. It is the tangibles, such as textbooks and written documentation, and the intangibles, expected norms of behaviour, that make up the professional practice of 'doing' physical geography. The rules defined by such practices could refer to the type of questions to be asked, the methods used to answer particular questions and the forms of analysis employed. These practices are not questioned, and once established by continual practice and repeated publication they become viewed as the way to undertake a particular type of study or the route to understand a particular phenomenon.

SOCIAL CONSTRUCTION AND PHYSICAL GEOGRAPHY

Recent discussions in geography have focused on the social construction of reality. Papers by Demeritt (1996, 2001a) in relation to climatology have put forward the argument that the reality studied by physical geographers is socially constructed. Countering this view, Schneider (2001) has suggested that such a strong social constructive view is detrimental to the subject. Specifically, this view highlights the relative nature of knowledge and denies the possibility of objective knowledge of the physical environment. Whilst social construction of reality is undeniable in the sense that physical geographers always work within social contexts, the strong 'sociology of science' view, that particular brands of social construction demand, paints a relativistic picture of knowledge that most, if not all, physical geographers feel uncomfortable with.

Discussion of social constructionism should begin, as Hacking (1999) notes, with assessing why you need to assert that something is socially constructed. Social construction in relation to social phenomena usually develops an argument concerning the negative nature of the phenomenon under investigation. Central to the argument is the seemingly natural nature of the phenomenon despite its social construction. There is no discussion of the reality of the phenomenon; its negative nature is to be negated by its social construction and, by implication, its non-existence in an objective reality.

Applying the same argument to physical geography implies that a physical and real basis to the entities and properties studied does not exist. This view of reality is at odds with other philosophies such as critical rationalism, critical realism and pragmatism which all assume the existence of an independent reality. Although these all also do not assume that this reality can be known directly, they do assume that it exists and that physical geography is striving to interpret it. Social constructivism, in its strongest sense, appears to deny this possibility or even encourage any attempt towards it.

It may be that the subject matter that is the focus of physical geography is different in nature from the subject matter of human geography. This is not to say that social constructivism, in the 'weak' sense identified above, cannot be applied to the subject matter but rather that the presence of an underlying reality that the subject matter refers to is not denied. As noted in Chapter 3, Hacking identified two types of kinds that can be the focus of study. Indifferent kinds are kinds or entities that do not react to the process of being classified. Interactive kinds are kinds or entities that do react and can respond to the process of being classified. Physical geography tends to deal with indifferent kinds or entities. They reflect the dialogue between the researcher and an independent reality.

The indifferent kinds do not necessarily accurately mirror reality, but they do arise from a social constructed, yet consistent, dialogue. It is this consistency and its agreed nature that provides the products of the dialogue with their objective power. Objective in this sense does not mean that the entities or kinds are 'real', only that they have been derived by a consistent process of dialogue and are treated as if real. The exact nature of the entity or kind may alter as the dialogue changes, but once defined the entities or kinds themselves influence the dialogue. In this sense it is pointless to begin to contrast 'real' as opposed to 'socially' constructed phenomena. The dichotomy is a false one. In effect, phenomena are both real and social at the same time.

Ethics in Physical Geography: Reflection Required

Ethical considerations in physical geography have not received much attention within the printed literature, although some debates have arisen at conferences. Scientific study in fields such as medicine and sociology have developed increasingly sophisticated means of assessing the ethical contents of their subject matter. Given that the focus of their research is on humans, this may seem an inappropriate place to look for guidelines about how to tackle ethical issues in physical geography. The basis that these fields have established does, however, provide information on the type of questions that should be considered. Within physical geography there are specific areas that should be questioned. First, at an abstract level the acceptance of an external, independent reality imposes a series of moral obligations on the scientific researcher. If the researcher is committed to uncovering, as far as they can, the operation of this reality, then there is an implicit obligation to 'truth'.

This is not as simple as it appears. Most physical geographers might assume that this means that the researchers should follow procedures diligently and record and revel in success and failure both the same. In other words, the researcher's work should reflect reality: the researcher holds a mirror up to reality and records the reflection without distortion. Hopefully, all of what has gone before illustrates the impossibility of such a naïve view of reality and its scientific study. Commitment to truth could be interpreted as ensuring that you derive data that reflect how reality is. Your view will be driven by theory and so truth, and these will become one and the same thing to you. Denying the theory is to deny the truth of how reality is. Reality may be interpreted to fit your theory. Ethically the researcher is committed to the truth, but other researchers may deny their version of the truth.

Commitment to the truth could be interpreted in another way. It could be viewed as commitment to use a specific approach to understanding reality and to abide by the outcomes of this approach. The centrality of falsification and testability in the philosophies outlined previously provides another means of reflecting reality. Researchers admit to the fallibility of their views of the world. Researchers submit their theories to rigorous testing in a dialogue with reality and with their co-researchers. Commitment to the truth then becomes commitment to this process of scientific scrutiny. It is through this method of rejection of theory that reality is to become knowable. Adherence to this process does, however, rely on an agreement between the researchers that the process of scrutiny is 'fair' and legitimate. Mechanisms for legitimating the process have evolved with the subject, from initial acceptance of the accounts of 'gentlemen' to the complicated peer-

review systems for academic papers and research projects. The development of legitimate pathways does, however, also constrain research that is acceptable or capable of judgement by the formats made available by these pathways. Second, physical geography is embedded within social networks as previously noted. These networks include moral ones that are variable in space and time. These networks may dictate what is an appropriate set of questions to ask, and what are appropriate methods of scientific investigation and appropriate forms of explanation. The early history of the study of the physical environment is littered with what seem now to be irrelevant moral constraints. Chorley et al. (1969) illustrated the need to retain the study of landscape development within a temporal framework that respected biblical events in the seventeenth and eighteenth centuries.

The development of a catastrophic view of landscape development then produced a reaction of strict uniformitarianism within geological and evolutionary thinking. Within contemporary society, of which physical geographers are a part, there is increasing discussion of the moral rights of the environment and its components. Issues of preservation and conservation of the physical environment impinge directly on the subject matter studied by physical geographers.

The indifferent entities and kinds that physical geographers study are part of the physical environment. Rivers, rocks and soils all contribute to the landscape, which has become a focus for moral concerns both in its own right and as the basis for supporting ecosystems and particular rare species. Concern that these entities are preserved for future generations as they are now, or conserved as dynamic entities, is the basis for this moral concern. A whole literature has developed concerning the moral and legal rights of future generations. Likewise, this concern has resulted in the development of sets of principles by which research activity is judged in terms of its impact upon the physical environment. Principles such as minimum impact and minimum tools are used within the US National Park Service to guide the assessment of the environmental impact of scientific research. These principles define the type of practical steps researchers need to build into their experimental designs to be granted permission to work within national parks. Principles such as minimum impact, for example, ask the researcher to consider how they reach the site. Do they take a car, what damage do they cause by their interference? and so on. The minimum-tool principle likewise requires the researcher to justify data collection using the proposed method. Issues such as environmental damage using a particular instrument relative to other instruments are put near the top of the agenda.

Each principle and its practical implications are designed with resource conservation in mind. The principles begin to question the assumptions researchers make about what are appropriate methods of study and bring to bear issues of general responsibility for actions that may have rarely entered into a researcher's consideration before. Third, a more general view could be taken of the ethical considerations within physical geography. The practice of physical geography involves a range of ethical choices, some of which are not viewed as choices at all but as established working practices. Recent discussions of the position of women within the earth sciences, for example, raise the issue of whether the practice of physical geography is inherently biased. The number of women, and the positions they occupy, could be viewed as being an indicator of gender-bias within physical geography. The development of an academic community requires that individuals acquire certain qualifications and become members of certain groups to be accepted in the field. Is selection of candidates for Ph.D.s based on gender lines?

Do decisions about the employment of academic staff get caught up with expectations of social worth and individuals' expected life choices (i.e. women get pregnant and look after the children, so are to be expected at work less)? You would hope to be able to say that in contemporary society these issues would not be a concern and that the answers to the above questions would all be an emphatic no! Suggestions such as physical fieldwork is based on a macho image of a fieldworker, and that the society to which they belong (e.g. a drinking and male-orientated culture) sometimes make it difficult to answer the above questions confidently. That these issues are rarely aired in physical geography does not mean they are not important.

In a similar vein there has recently been concern for the 'invisible' researchers within geography. The Research Assessment Exercise in the UK has thrown into sharp relief the role and lack of career structure for post-doctorate researchers and even Ph.D. students. The trends in the employment of this group could be viewed in the same light as the general trend towards casualization of the labour force as a whole with all the associated problems. Academics, often the employers of these individuals, have been relatively slow to appreciate and develop strategies for career development. Often academics working with short-term contract researchers do not see it as part of their job to think about their employees. Often the employees are unaware of what their rights are. Likewise, comments are usually made about the increased flexibility of work practices for researchers by full-time academics.

The dialogue between the groups, if indeed there are identifiable groupings, is one of an unequal power relationship. Oddly, academics have rarely researched this personal and close relationship between themselves and their researchers. Increasingly, in the UK at least, as research funds become concentrated in a few 'top' institutions, the development of academic careers will be controlled by fewer and fewer individuals. The ethical implications of this have rarely if ever been discussed. The above discussions illustrate the important role that issues of responsibility play in developing an ethical code for research in physical geography. Although policy instruments such as minimum impact and minimum tools impose a framework of compliance upon researchers, they can also be viewed as practical and codified practices for environmental responsibility. The development of an ethical framework can be imposed from above, as in this case, but it is more relevant to ask why the framework has to be imposed in the first place. Questions of moral uncertainty may be part of the answer, but it is likely that unaddressed issues of researcher responsibility are also part of the problem. Scientific research tends to focus on a particular problem.

As the book has highlighted, and as others have commented, science is the art of the possible. Researchers break questions down into manageable chunks and answer what they believe they can using the tools that they have learnt to use. Questioning of the questions being asked and the impact of their investigations upon the state of the physical environment for future generations is not a question directly involved in their framing of research. Particularly where research is focused around the short-term research grant with its associated promotional rewards, individual justification for an action may outweigh potential ethical concerns if they are even articulated. Development of an ethical framework needs to be developed within the individual researcher. A researcher may claim they have an ethical commitment to discover the 'truth' of reality, but what do they do if the method they intend to employ causes irreversible damage to the environment they are studying? Which ethical concern has the greater moral imperative? Development of such an individual ethical framework will require a lot more work at the level of individuals and their training, as well as being translated as a payback in career success. If the context of the individual researcher is not given as much weight as the ethical argument, then any ethical framework will founder on the grim personal practicalities of the individual.

Although great consideration has been given to the ethics of research in the physical environment, the same consideration should be given to ethics in the process of research. Increasingly, research is a group or team activity rather than the solitary undertakings of a lone scholar. Similarly, with the advent of the Research Assessment Exercise in the UK, as well as the traditional peer-review system for publications, success of research activity is dependent upon a community. How individuals deal with other individuals in this research process is as much the sphere of ethical considerations as is the output from the process, the research itself.

Once again this sort of issue is something that academics have not had to deal with and often it does not even register as part of their ethical framework. Within the UK, for example, there has recently been a concern over the casualization of researchers, almost the development of a research underclass. The status of these individuals in the research process is often left to the vagaries of individual research team leaders. The responsibility of employer to employee and vice versa are part of this relationship, but one which academics are rarely trained for or necessarily have an ethical framework in which to assess. It is telling that a number of universities have developed rules for inclusion and precedence of authorship on academic papers involving such teams. These raise issues of unequal power relationships within the research process. The inequality of these relationships requires a recognition and consideration that has been lacking in the recent past.

PHYSICAL AND HUMAN GEOGRAPHY: DIVISION OR INTEGRATION

Recent discussions over the nature of geography often refer to the human/physical divide in the subject. Despite protests and pleading to the contrary, the divide is often seen as insurmountable, with some even claiming that geography has become effectively split into two specialisms (some even carry the subdivision amongst specialisms further). Massey (1999) tried to tease out the commonalities between the two parts of geography via a consideration of the role of entities. Raper and Livingstone (1995) confirm the potential for discussion, as did the ensuing printed debate. Despite these overtures across the divide, there seems to be little real movement of positions so far. This debate about the possibility of integrating physical and human geography is not a new one. A special issue of the Transactions of the Institute of British Geographers in 1986 provided a series of short papers on different geographers' opinions on the likelihood of such common study. These concerns and hopes have also been expressed further back in the century, as in Man's Role in Changing the Face of the Earth (Thomas, 1956). Several geographers have explicitly been concerned with developing concepts that would provide this integration, such as Carl Sauer and the concept of landscape (1956). By the end of the twentieth century, the debate still continued without resolution.

Keeping the two sides apart are centrifugal forces that vary from place to place, but which collectively are likely to prevent the integration wished for by many. Academics, as noted above, are only human. Career paths and funding requirements mean that people will pursue courses of action profitable to them in their particular context. If funding councils do not recognize an integrative physical/human geography then why pursue it? There is no money in it; there are no rewards. This may sound cynical, but an academic life is short, the period available to make an impact even shorter, so why waste it in pursuit of an unattainable goal? There is also the feeling amongst key researchers on both sides that what they study is different from that being studied on the other side. The subject matter of human and physical geography is seen as fundamentally different. Hacking's (1999) distinction between indifferent and interactive kinds may be taken by some to be a good description of the differences between physical and human geography respectively. This may be simplifying the matter, but it does have an appeal in explaining the continued divide. If the subject matter is different at a fundamental level then the techniques employed to study one kind cannot necessarily be easily transferred to the study of the other kind. This means that a reductionist approach will not be of any use in a post-modern analysis of asylum seekers. Likewise, what could a post-modernist say that is meaningful to the study of gravel-bedded rivers? Acceptance of such a key divide will justify the lack of communication between the camps.

With such entrenched and well-funded camps is there any point in pursuing the issue of integrating physical and human geography? Maybe, unsurprisingly, my reply is yes! Although some may regard any integration as superficial and part of a propaganda drive to maintain a university subject that has run its course, I still believe that there is room for an integrative approach. The methodologies of human geography can be applied to the study of the practice of physical geography. Collier and Inkpen (2002, 2003), somewhat simplistically, began to analyse how surveying at the RGS in the nineteenth century was influenced in its practice by the context of the period. Livingstone (1984, 1987) has analysed Shaler's work in both eugenics and geology in the context of the development of Darwinian concepts at the time. Such historic studies can help us to see how seemingly objective facts and concepts are actually constructed by individuals in societies and how the nature of 'objectivity' is affected by these contexts.

Likely to be of more significance to current physical geography is the increasing recognition (stating the obvious maybe) that it is difficult to separate human and physical environments. The interpenetration of the physical and human means that it is difficult to justify that processes of environmental change are purely physical or that social structures rely solely upon human processes. Maintaining that one realm does not influence the other would be to view each as meshing together at increasingly smaller scales, but never actually overlapping or merging. Gould (2001) provides a similar image of the relationship between science and religion, labelling it as 'fractal' as the two spheres mixed, but never merged no matter the scale of resolution they were observed at. For human geography I mean the more humanistic approaches that seem to address different questions and a different reality from the 'scientific' approaches.

The seemingly clear divide between the two becomes messy when closely examined in terms of, say, the purpose of study, the questions asked, the entities and processes identified. The two, however, never mix, their realities remain different but interwoven. The two never merged as each answered questions about a specific realm: science, what reality was like; religion, how we ought to live. One could not answer questions about the other. Keeping human and physical geography separate implies the same key distinction. Interestingly, however, this separatist view would not actually deny the importance of either branch of geography; it would merely accept that human and physical questions were distinct and separate, but equally as valid to ask in the first place. This view would enable an uneasy truce to prevail in conflict over the integration or division of the subject.

An important consideration that has rarely been considered is that both human and physical geography are not homogeneous masses. There are distinct approaches within both with differing methodologies and views of reality. Arguments for separation, and even some for integration, have usually focused upon the extreme members of both realms. The argument seems to run that if these end members cannot be integrated then none of physical and human geography can. It may be that these end members follow Gould's fractal description, but other parts of human and physical geography may be easier or rather more appropriate to integrate. The approaches and topics that ask the same questions about reality or can inform the questions that each subdiscipline asks may be more appropriate candidates for integration. As environmental degradation continues, this may be an area that of necessity becomes a concern for geography as a subject for both its relevance and survival.

SUMMARY

Physical geographers work within different societies at different scales. All are members of different networks and may behave in different, even contradictory, manners, depending on the network being considered. Individuals should not be seen, however, as automatons responding blindly or consistently to their context. Context is mediated by individuals in a complicated and often unique manner. Recently, ethical concerns have become an important consideration in physical geography, whether forces upon researchers through instruments such as minimum impact or from within by moral considerations. There is little in terms of a coherent and explicit professional code of conduct for physical geographers to follow. Ethical obligations to the truth, to the physical environment being studied, to other researchers and to the individual are all aspects of the research process that are not usually considered in detail.

9

MAPS, REMOTE SENSING, AND GIS

INTRODUCTION

A map can be simply defined as a graphic representation of the real world. This representation is always an abstraction of reality. Because of the infinite nature of our Universe it is impossible to capture all of the complexity found in the real world. For example, topographic maps abstract the three-dimensional real world at a reduced scale on a two-dimensional plane of paper.

Figure: The following specialized weather map displays the surface location of pressure centers and fronts for Saturday, November 27, 1999 over a portion of North America.

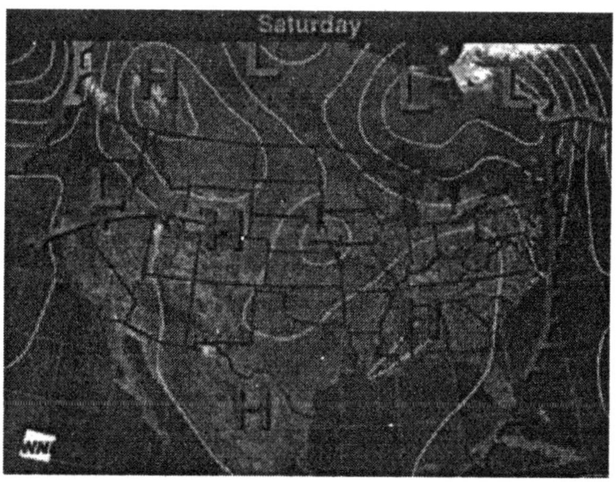

Maps are used to display both cultural and physical features of the environment. Standard topographic maps show a variety of information including roads, land-use classification, elevation, rivers and other water bodies, political boundaries, and the identification of houses and other types of buildings. Some maps are created with very specific goals in mind. Figure 2a-1 displays a weather map showing the location of low and high pressure centers and fronts over most of North America. The intended purpose of this map is considerably more specialized than a topographic map.

The art of map construction is called cartography. People who work in this field of knowledge are called cartographers. The construction and use of maps has a long history. Some academics believe that the earliest maps date back to the fifth or sixth century BC. Even in these early maps, the main goal of this tool was to communicate information. Early maps were quite subjective in their presentation of spatial information. Maps became more objective with the dawn of Western science. The application of scientific method into cartography made maps more ordered and accurate. Today, the art of map making is quite a sophisticated science employing methods from cartography, engineering, computer science, mathematics, and psychology.

Cartographers classify maps into two broad categories: reference maps and thematic maps. Reference maps normally show natural and human-made objects from the geographical environment with an emphasis on location. Examples of general reference maps include maps found in atlases and topographic maps. Thematic maps are used to display the geographical distribution of one phenomenon or the spatial associations that occur between a number of phenomena.

MAP PROJECTION

The shape of the Earth's surface can be described as an ellipsoid. An ellipsoid is a three-dimensional shape that departs slightly from a purely spherical form. The Earth takes this form because rotation causes the region near the equator to bulge outward to space. The angular motion caused by the Earth spinning on its axis also forces the polar regions on the globe to be somewhat flattened.

Representing the true shape of the Earth's surface on a map creates some problems, especially when this depiction is illustrated on a two-dimensional surface. To overcome these problems, cartographers have developed a number of standardized transformation processes for the creation of two-dimensional maps. All of these transformation processes create some type of distortion artifact. The nature of this distortion is related to how the transformation process modifies specific geographic properties of the map. Some of the geographic properties affected by projection distortion include: distance; area; straight line direction between points on the Earth; and the bearing of cardinal points from locations on our planet.

The illustrations below show some of the common map projections used today. The first two-dimensional projection shows the Earth's surface as viewed from space. This orthographic projection distorts distance, shape, and the size of areas. Another serious limitation of this projection is that only a portion of the Earth's surface can be viewed at any one time.

The second illustration displays a Mercator projection of the Earth. On a Mercator projection, the north-south scale increases from the equator at the same rate as the corresponding east-west scale.

As a result of this feature, angles drawn on this type of map are correct. Distortion on a Mercator map increases at an increasing rate as one moves toward higher latitudes. Mercator maps are used in navigation because a line drawn between two points of the Earth has true direction. However, this line may not represent the shortest distance between these points.

Figure: Earth as observed from a vantage point in space. This orthographic projection of the Earth's surface creates a two-dimensional representation of a three-dimensional surface. The orthographic projection distorts distance, shape, and the size of areas.

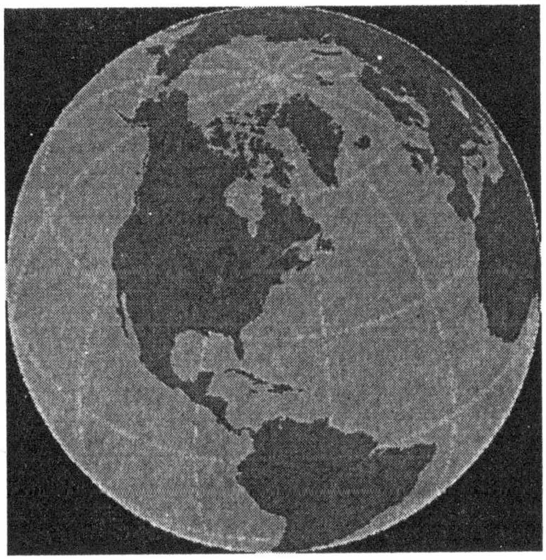

Figure: Mercator map projection. The Mercator projection is one of the most common systems in use today. It was specifically designed for nautical navigation.

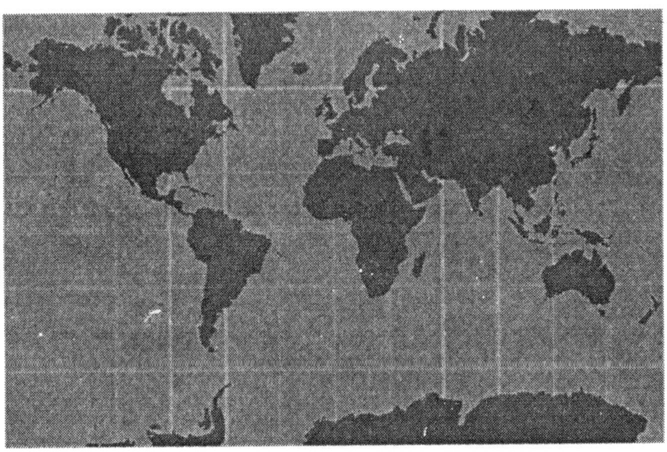

The Gall-Peters projection was developed to correct some of the distortion found in the Mercator system. The Mercator projection causes area to be gradually distorted from the equator to the poles. This distortion makes middle and high latitude countries to be bigger than they are in reality. The Gall-Peters projection corrects this distortion making the area occupied by the world's nations more comparable.

Figure: Gall-Peters projection. The Gall-Peters projection corrects the distortion of area common in Mercator maps. As a result, it removes the bias in Mercator maps that draws low latitude countries as being smaller than nations in middle and high latitudes. This projection has been officially adopted by a number of United Nations organizations.

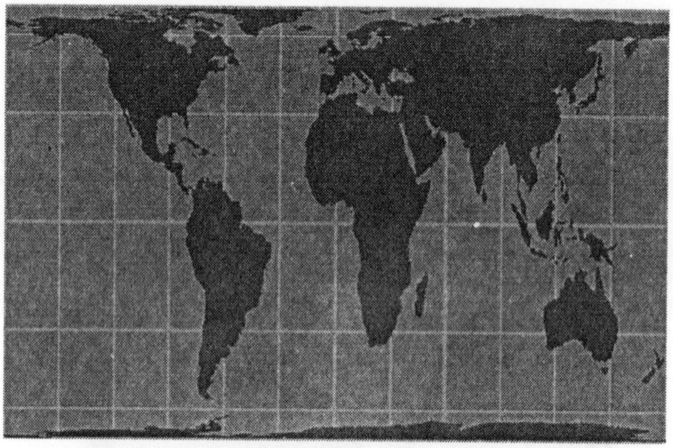

The Miller Cylindrical projection is another common two-dimensional map used to represent the entire Earth in a rectangular area. In this project, the Earth is mathematically projected onto a cylinder tangent at the equator.

This projection in then unrolled to produce a flat two-dimensional representation of the Earth's surface. This projection reduces some of the scale exaggeration present in the Mercator map. However, the Miller Cylindrical projection describes shapes and areas with considerable distortion and directions are true only along the equator.

Figure: The Miller Cylindrical projection.

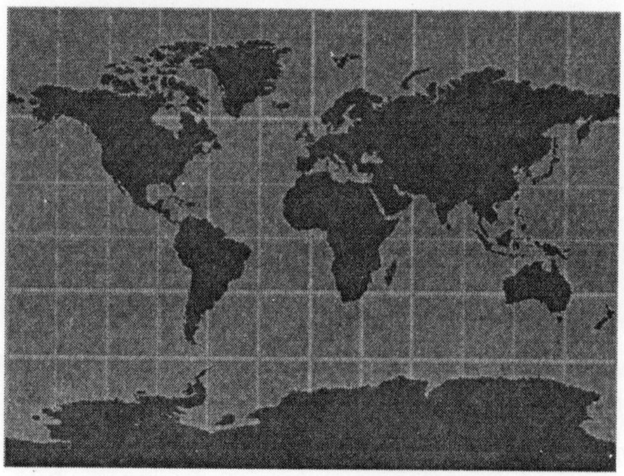

Figure below displays the Robinson projection. This projection was developed to show the entire Earth with less distortion of area. However, this feature requires a tradeoff in terms of inaccurate map direction and distance.

Figure: Robinson's projection. This projection is common in maps that require somewhat accurate representation of area. This map projection was originally developed for Rand McNally and Company in 1961.

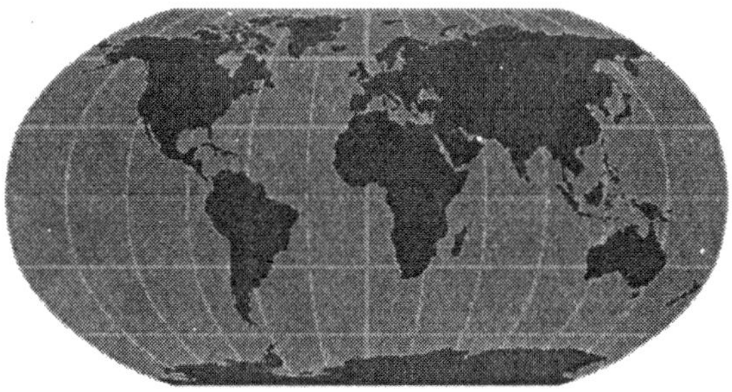

The Mollweide projection improves on the Robinson projection and has less area distortion. The final projection presented presents areas on a map that are proportional to the same areas on the actual surface of the Earth. However, this Sinusoidal Equal-Area projection suffers from distance, shape, and direction distortions.

Figure: Mollweide projection. On this projection the only parallels (line of latitude) drawn of true length are 40° 40' North and South. From the equator to 40° 40' North and South the east west scale is illustrated too small. From the poles to 40° 40' North and South the east-west scale is illustrated too large.

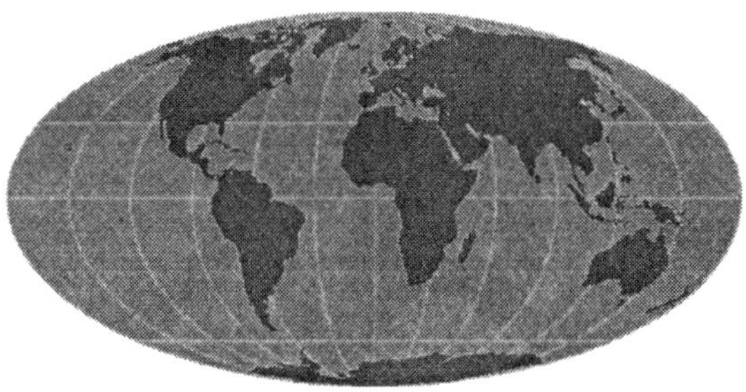

Figure: Sinusoidal Equal-Area projection.

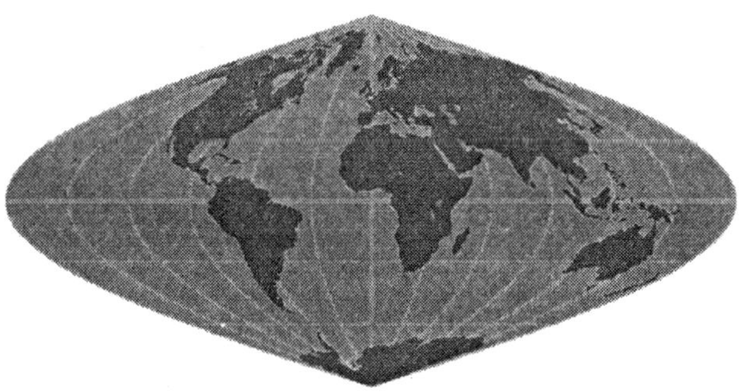

Map Scale

Maps are rarely drawn at the same scale as the real world. Most maps are made at a scale that is much smaller than the area of the actual surface being depicted. The amount of reduction that has taken place is normally identified somewhere on the map. This measurement is commonly referred to as the map scale. Conceptually, we can think of map scale as the ratio between the distance between any two points on the map compared to the actual ground distance represented. This concept can also be expressed mathematically as:

$$\text{Map Scale} = \frac{\text{Map Distance}}{\text{Earth Distance}}$$

On most maps, the map scale is represented by a simple fraction or ratio. This type of description of a map's scale is called a representative fraction. For example, a map where one unit (centimeter, meter, inch, kilometer, etc.) on the illustration represents 1,000,000 of these same units on the actual surface of the Earth would have a representative fraction of 1/1,000,000 (fraction) or 1:1,000,000 (ratio). Of these mathematical representations of scale, the ratio form is most commonly found on maps.

Scale can also be described on a map by a verbal statement. For example, 1:1,000,000 could be verbally described as "1 centimeter on the map equals 10 kilometers on the Earth's surface" or "1 inch represents approximately 16 miles".

Most maps also use graphic scale to describe the distance relationships between the map and the real world. In a graphic scale, an illustration is used to depict distances on the map in common units of measurement. Graphic scales are quite useful because they can be used to measure distances on a map quickly.

Figure: The following graphic scale was drawn for map with a scale of 1:250,000. In the illustration distances in miles and kilometers are graphically shown.

Maps are often described, in a relative sense, as being either small scale or large scale. Figure 2a-10 helps to explain this concept. In Figure below, we have maps representing an area of the world at scales of 1:100,000, 1:50,000, and 1:25,000. Of this group, the map drawn at 1:100,000 has the smallest scale relative to the other two maps. The map with the largest scale is map C which is drawn at a scale of 1:25,000.

Figure: The following three illustrations describe the relationship between map scale and the size of the ground area shown at three different map scales. The map on the far left has the smallest scale, while the map on the far right has the largest scale. Note what happens to the amount of area represented on the maps when the scale is changed. A doubling of the scale (1:100,000 to 1:50,000 and 1:50,000 to 1:25,000) causes the area shown on the map to be reduced to 25% or one-quarter.

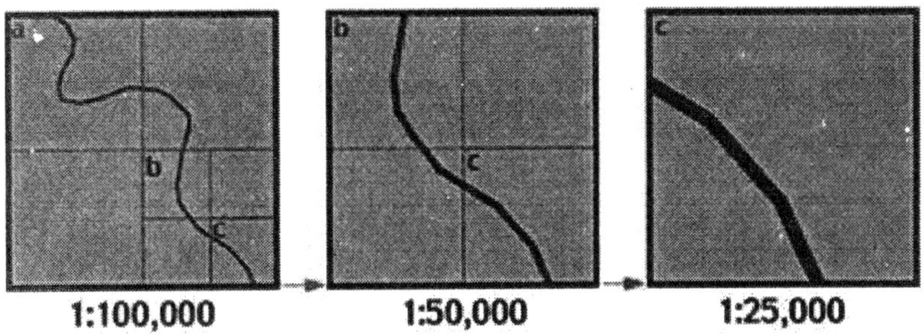

LOCATION, DISTANCE, AND DIRECTION ON MAPS

Location on Maps

Most maps allow us to specify the location of points on the Earth's surface using a coordinate system. For a two-dimensional map, this coordinate system can use simple geometric relationships between the perpendicular axes on a grid system to define spatial location. Figure below illustrates how the location of a point can be defined on a coordinate system.

Figure: A grid coordinate system defines the location of points from the distance traveled along two perpendicular axes from some stated origin. In the example above, the two axes are labeled X and Y. The origin is located in the lower left hand corner. Unit distance traveled along each axis from the origin is shown. In this coordinate system, the value associated with the X-axis is given first, following by the value assigned from the Y-axis. The location represented by the star has the coordinates 7 (X-axis), 4 (Y-axis).

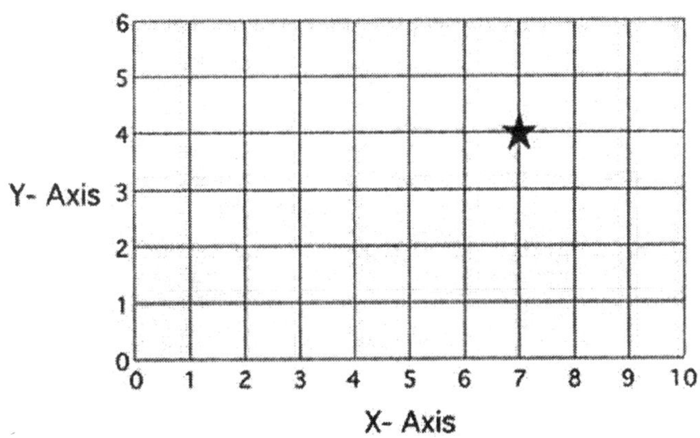

Two types of coordinate systems are currently in general use in geography: the geographical coordinate system and the rectangular (also called Cartesian) coordinate system.

Geographical Coordinate System

The geographical coordinate system measures location from only two values, despite the fact that the locations are described for a three-dimensional surface. The two values used to define location are both measured relative to the polar axis of the Earth. The two measures used in the geographic coordinate system are called latitude and longitude.

Figure: Lines of latitude or parallels are drawn parallel to the equator (shown in red) as circles that span the Earth's surface. These parallels are measure in degrees (°). There are 90 angular degrees of latitude from the equator to each of the poles. The equator has an assigned value of 0°. Measurements of latitude are also defined as being either north or south of equator to distinguish the hemisphere of their location. Lines of longitude or meridians are circular arcs that meet at the poles. There are 180° of longitude either side of a starting meridian which is known the Prime Meridian. The Prime Meridian has a designated value of 0°. Measurements of longitude are also defined as being either west or east of the Prime Meridian.

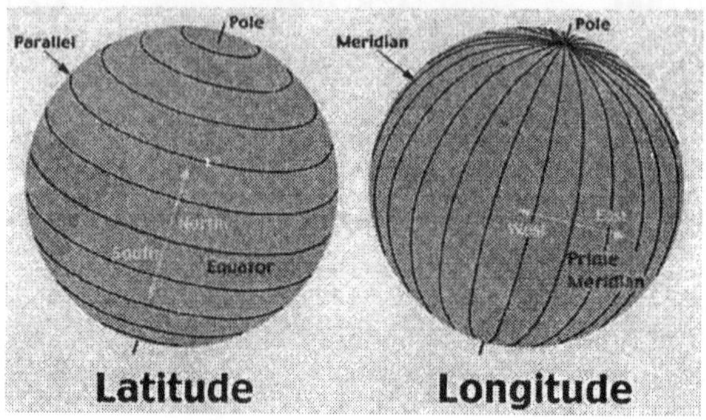

Latitude measures the north-south position of locations on the Earth's surface relative to a point found at the center of the Earth. This central point is also located on the Earth's rotational or polar axis. The equator is the starting point for the measurement of latitude. The equator has a value of zero degrees. A line of latitude or parallel of 30° North has an angle that is 30° north of the plane represented by the equator. The maximum value that latitude can attain is either 90° North or South. These lines of latitude run parallel to the rotational axis of the Earth.

Figure: Measurement of latitude and longitude relative to the equator and the Prime Meridian and the Earth's rotational or polar axis.

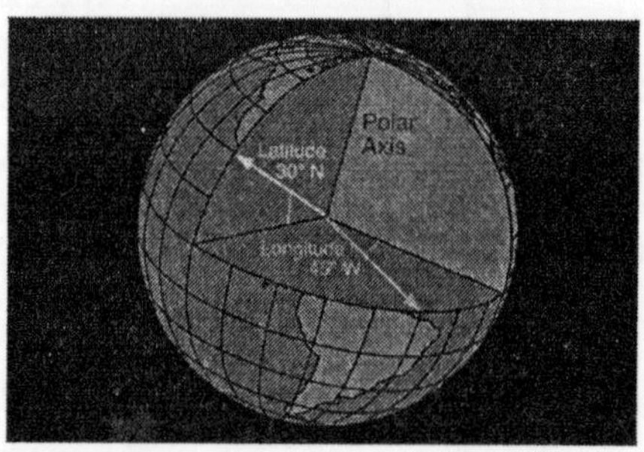

Longitude measures the west-east position of locations on the Earth's surface relative to a circular arc called the Prime Meridian. The position of the Prime Meridian was determined by international agreement to be in-line with the location of the former astronomical observatory at Greenwich, England. Because the Earth's circumference is similar to circle, it was decided to measure longitude in degrees. The number of degrees found in a circle is 360. The Prime Meridian has a value of zero degrees. A line of longitude or meridian of 45° West has an angle that is 45° west of the plane represented by the Prime Meridian. The maximum value that a meridian of longitude can have is 180° which is the distance halfway around a circle. This meridian is called the International Date Line. Designations of west and east are used to distinguish where a location is found relative to the Prime Meridian. For example, all of the locations in North America have a longitude that is designated west.

UNIVERSAL TRANSVERSE MERCATOR SYSTEM (UTM)

Another commonly used method to describe location on the Earth is the Universal Transverse Mercator (UTM) grid system. This rectangular coordinate system is metric, incorporating the meter as its basic unit of measurement. UTM also uses the Transverse Mercator projection system to model the Earth's spherical surface onto a two-dimensional plane.

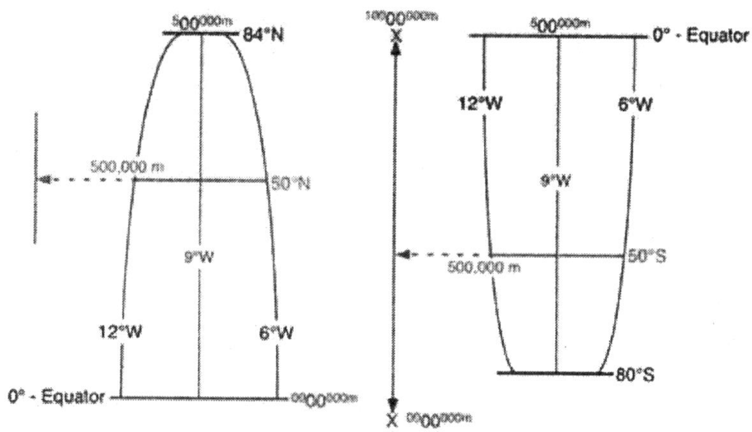

Figure: *The following illustration describes the characteristics of the UTM zone "29" found between 12 to 6° West longitude. Note that the zone has been split into two halves. The half on the left represents the area found in the Northern Hemisphere. The Southern Hemisphere is located on the right. The blue line represents the central meridian for this zone. Locations measurements for this zone are calculated relative to a false origin. In the Northern Hemisphere, this origin is located 500,000 meters west of the equator. The Southern Hemisphere UTM measurements are determined relative to a origin located at 10,000,000 meters south and 500,000 meters west of the equator and central meridian, respectively.*

The UTM system divides the world's surface into 60 - six degree longitude wide zones that run north-south. These zones start at the International Date Line and are successively numbered in an eastward direction. Each zone stretches from 84° North to 80° South. In the center of each of these zones is a central meridian. Location is measured in these zones from a false origin which is determined relative to the intersection of the equator and the central meridian for each zone. For locations in the Northern Hemisphere, the false origin is 500,000 meters west of the central meridian on the equator. Coordinate measurements of location in the Northern Hemisphere using the UTM system are made relative to this point in meters in eastings (longitudinal distance) and northings (latitudinal distance). The point defined by the intersection of 50° North and 9° West would have a UTM coordinate of Zone 29, 500000 meters east (E), 5538630 meters north (N). In the Southern Hemisphere, the origin is 10,000,000 meters south and 500,000 meters west of the equator and central meridian, respectively. The location found at 50° South and 9° West would have a UTM coordinate of Zone 29, 500000 meters E, 4461369 meters N (remember that northing in the Southern Hemisphere is measured from 10,000,000 meters south of the equator.

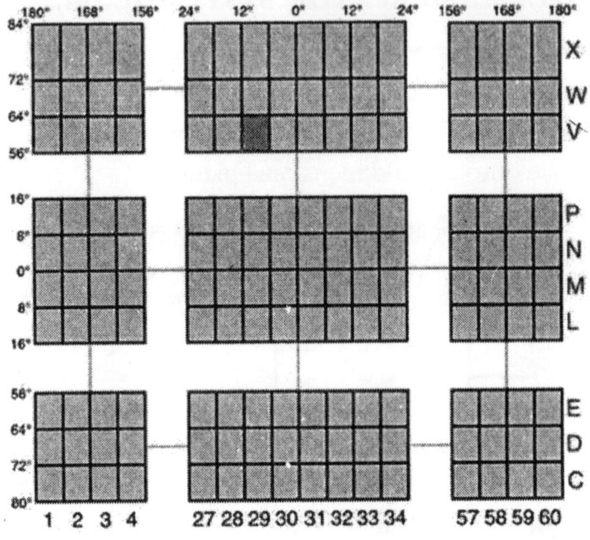

Figure: *The UTM system also uses a grid system to break the Earth up into 1200 quadrilaterals. To keep the illustration manageable, most of these zones have been excluded. Designation of each quadrilaterals is accomplished with a number-letter system. Along the horizontal bottom, the six degree longitude wide zones are numbered, starting at 180° West longitude, from 1 to 60. The twenty vertical rows are assigned letters C to X with I and O excluded. The letter, C, begins at 80° South latitude. Note that the rows are 8 degrees of latitude wide, except for the last row X which is 12 degrees wide. According to the reference system, the bright green quadrilateral has the grid reference 29V (note that in this system west-east coordinate is given first, followed by the south-north coordinate). This grid zone is found between 56 and 64° North latitude and 6 and 12° West longitude.*

The UTM system has been modified to make measurements less confusing. In this modification, the six degree wide zones are divided into smaller pieces or quadrilaterals that are eight degrees of latitude tall. Each of these rows is labeled, starting at 80° South, with the letters C to X consecutively with I and O being omitted. The last row X differs from the other rows and extends from 72 to

84° North latitude (twelve degrees tall). Each of the quadrilaterals or grid zones are identified by their number/letter designation. In total, 1200 quadrilaterals are defined in the UTM system.

The quadrilateral system allows us to further define location using the UTM system. For the location 50° North and 9° West, the UTM coordinate can now be expressed as Grid Zone 29U, 500000 meters E, 5538630 meters N.

Each UTM quadrilateral is further subdivided into a number of 100,000 by 100,000 meter zones. These subdivisions are coded by a system of letter combinations where the same two-letter combination is not repeated within 18 degrees of latitude and longitude. Within each of the 100,000 meter squares one can specify location to one-meter accuracy using a 5 digit eastings and northings reference system.

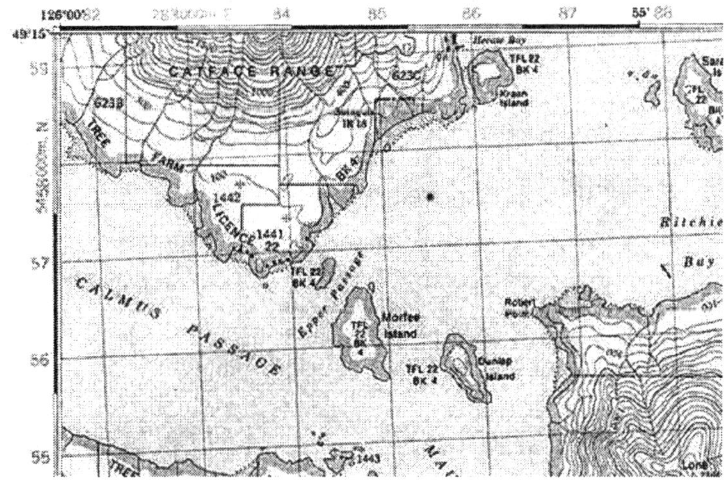

Figure: *The top left hand corner the "Tofino" 1:50,000 National Topographic Series of Canada map is shown above. The blue lines and associated numbers on the map margin are used to determine location by way of the UTM grid system. Abbreviated UTM 1,000-meter values or principle digits are shown by numbers on the map margin that vary from 0 to 100 (100 is actually given the value 00). In each of the corners of the map, two of the principle digits are expressed in their full UTM coordinate form. On the image we can see 283000 m E. and 5458000 m N. The red dot is found in the center of the grid defined by principle numbers 85 to 86 easting and 57 to 58 northing. A more complete UTM grid reference for this location would be 285500 m E. and 5457500 m N. Information found on the map margin also tells us (not shown) that the area displayed is in Grid Zone 10U and the 100,000 m squares BK and CK are located on this map.*

The UTM grid system is displayed on all United States Geological Survey (USGS) and National Topographic Series (NTS) of Canada maps. On USGS 7.5-minute quadrangle maps (1:24,000 scale), 15-minute quadrangle maps (1:50,000, 1:62,500, and standard-edition 1:63,360 scales), and Canadian 1:50,000 maps the UTM grid lines are drawn at intervals of 1,000 meters, and are shown either with blue ticks at the edge of the map or by full blue grid lines. On USGS maps at 1:100,000 and

1:250,000 scale and Canadian 1:250,000 scale maps a full UTM grid is shown at intervals of 10,000 meters. Figure below7 describes how the UTM grid system can be used to determine location on a 1:50,000 National Topographic Series of Canada map.

Distance on Maps

In previous section, w e have learned that depicting the Earth's three-dimensional surface on a two-dimensional map creates a number of distortions that involve distance, area, and direction. It is possible to create maps that are somewhat equidistance. However, even these types of maps have some form of distance distortion. Equidistance maps can only control distortion along either lines of latitude or lines of longitude. Distance is often correct on equidistance maps only in the direction of latitude.

On a map that has a large scale, 1:125,000 or larger, distance distortion is usually insignificant. An example of a large-scale map is a standard topographic map. On these maps measuring straight line distance is simple. Distance is first measured on the map using a ruler. This measurement is then converted into a real world distance using the map's scale. For example, if we measured a distance of 10 centimeters on a map that had a scale of 1:10,000, we would multiply 10 (distance) by 10,000 (scale). Thus, the actual distance in the real world would be 100,000 centimeters.

Measuring distance along map features that are not straight is a little more difficult. One technique that can be employed for this task is to use a number of straight-line segments. The accuracy of this method is dependent on the number of straight-line segments used. Another method for measuring curvilinear map distances is to use a mechanical device called an opisometer. This device uses a small rotating wheel that records the distance traveled. The recorded distance is measured by this device either in centimeters or inches.

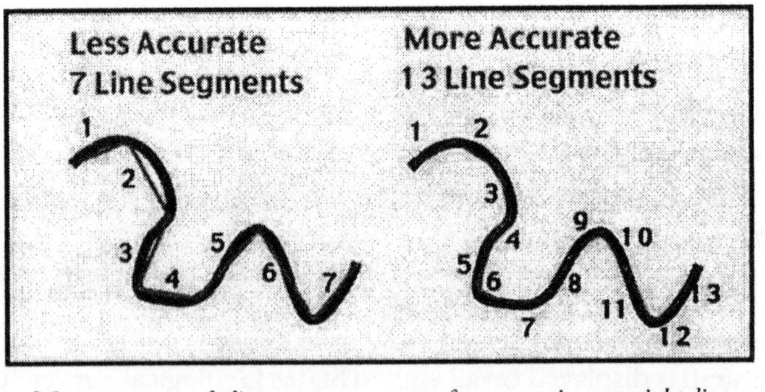

Figure: *Measurement of distance on a map feature using straight-line segments.*

Direction on Maps

Like distance, direction is difficult to measure on maps because of the distortion produced by projection systems. However, this distortion is quite small on maps with scales larger than 1:125,000.

Direction is usually measured relative to the location of North or South Pole. Directions determined from these locations are said to be relative to True North or True South.

The magnetic poles can also be used to measure direction. However, these points on the Earth are located in spatially different spots from the geographic North and South Pole. The North Magnetic Pole is located at 78.3° North, 104.0° West near Ellef Ringnes Island, Canada. In the Southern Hemisphere, the South Magnetic Pole is located in Commonwealth Day, Antarctica and has a geographical location of 65° South, 139° East. The magnetic poles are also not fixed overtime and shift their spatial position overtime.

Topographic maps normally have a declination diagram drawn on them. On Northern Hemisphere maps, declination diagrams describe the angular difference between Magnetic North and True North. On the map, the angle of True North is parallel to the depicted lines of longitude. Declination diagrams also show the direction of Grid North. Grid North is an angle that is parallel to the easting lines found on the Universal Transverse Mercator (UTM) grid system.

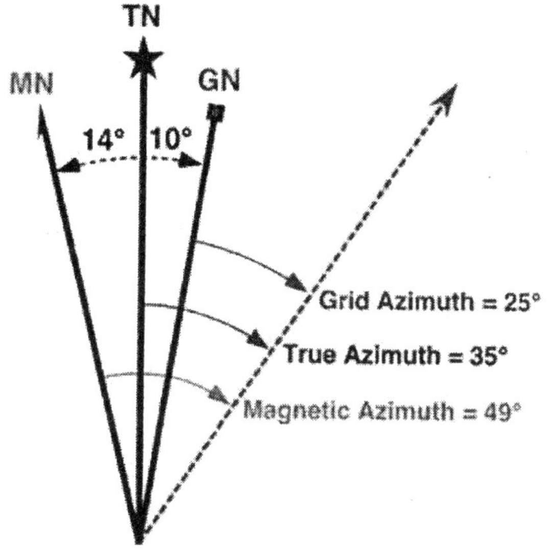

Figure: *This declination diagram describes the angular difference between Grid, True, and Magnetic North. This illustration also shows how angles are measured relative grid, true, and magnetic azimuth.*

In the field, the direction of features is often determined by a magnetic compass which measures angles relative to Magnetic North. Using the declination diagram found on a map, individuals can convert their field measures of magnetic direction into directions that are relative to either Grid or True North. Compass directions can be described by using either the azimuth system or the bearing system. The azimuth system calculates direction in degrees of a full circle. A full circle has 360 degrees. In the azimuth system, north has a direction of either the 0 or 360°. East and west have an azimuth of 90° and 270°, respectively. Due south has an azimuth of 180°.

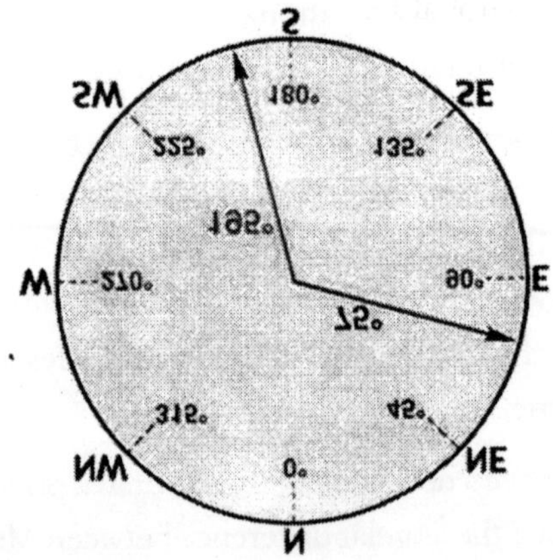

Figure: *Azimuth system for measuring direction is based on the 360 degrees found in a full circle. The illustration shows the angles associated with the major cardinal points of the compass. Note that angles are determined clockwise from north.*

The bearing system divides direction into four quadrants of 90 degrees. In this system, north and south are the dominant directions. Measurements are determined in degrees from one of these directions. The measurement of two angles based on this system are described in Figure below.

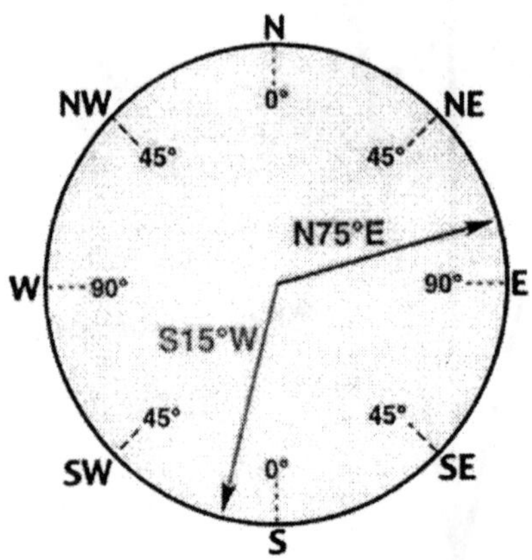

Figure: *The bearing system uses four quadrants of 90 degrees to measure direction. The illustration shows two direction measurements. These measurements are made relative to either north or south. North and south are given the measurement 0 degrees. East and west have a value of 90 degrees. The first measurement (green) is found in the north - east quadrant. As a result, its measurement is north 75 degrees to the east or N75°E. The first measurement (orange) is found in the south - west quadrant. Its measurement is south 15 degrees to the west or S15°W.*

Global Positioning Systems

Determination of location in field conditions was once a difficult task. In most cases, it required the use of a topographic map and landscape features to estimate location. However, technology has now made this task very simple. Global Positioning Systems (GPS) can calculate one's location to an accuracy of about 30-meters. These systems consist of two parts: a GPS receiver and a network of many satellites. Radio transmissions from the satellites are broadcasted continually. The GPS receiver picks up these broadcasts and through triangulation calculates the altitude and spatial position of the receiving unit. A minimum of three satellite is required for triangulation.

Map Location and Time Zones

Before the late nineteenth century, time keeping was essentially a local phenomenon. Each town would set their clocks according to the motions of the Sun. Noon was defined as the time when the Sun reached its maximum altitude above the horizon. Cities and towns would assign a clockmaker to calibrate a town clock to these solar motions. This town clock would then represent "official" time and the citizens would set their watches and clocks accordingly.

The later half of the nineteenth century was time of increased movement of humans. In the United States and Canada, large numbers of people were moving west and settlements in these areas began expanding rapidly. To support these new settlements, railroads moved people and resources between the various cities and towns. However, because of the nature of how local time was kept, the railroads experience major problems in constructing timetables for the various stops. Timetables could only become more efficient if the towns and cities adopted some type of standard method of keeping time. In 1878, Canadian Sir Sanford Fleming suggested a system of worldwide time zones that would simplify the keeping of time across the Earth. Fleming proposed that the globe be divided into 24 time zones, each 15 degrees of longitude in width. Since the world rotates once every 24 hours on its axis and there are 360 degrees of longitude, each hour of Earth rotation represents 15 degrees of longitude.

Railroad companies in Canada and the United States began using Fleming's time zones in 1883. In 1884, an International Prime Meridian Conference was held in Washington D.C. to adopted the standardize method of time keeping and determined the location of the Prime Meridian. Conference members agreed that the longitude of Greenwich, England would become zero degrees longitude and established the 24 time zones relative to the Prime Meridian. It was also proposed that the measurement of time on the Earth would be made relative to the astronomical measurements at the Royal Observatory at Greenwich. This time standard was called Greenwich Mean Time (GMT).

Today, many nations operate on variations of the time zones suggested by Sir Fleming. Figure below describes the various time zones currently used on the Earth. In this system, time in the various zones is measured relative the Coordinated Universal Time (UTC) standard at the Prime Meridian. Coordinated Universal Time became the standard legal reference of time all over the world in 1972. UTC is determined from six primary atomic clocks that are coordinated by the International Bureau of Weights and Measures (BIPM) located in France. The numbers located at the bottom of Figure below indicate how many hours each zone is earlier (negative sign) or later (positive sign) than the Coordinated Universal Time standard. Also note that national boundaries and political matters influence the shape of the time zone boundaries. For example, China uses a single time zone (eight hours ahead of Coordinated Universal Time) instead of five different time zones.

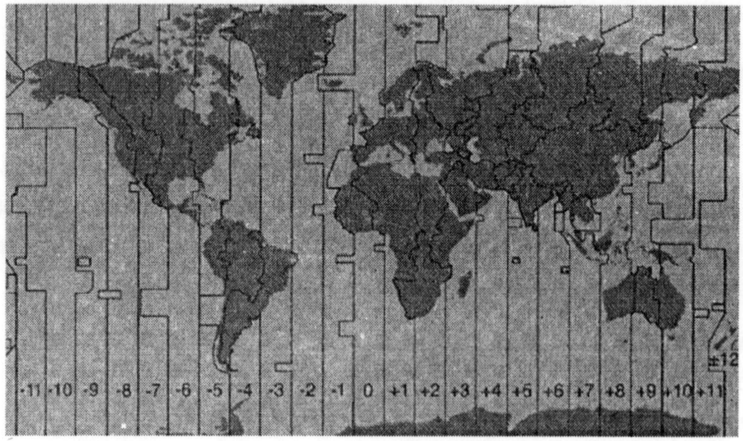

Figure: *Modern standard times zones as measured relative to Coordinated Universal Time. The numbers located at the bottom indicate how many hours each zone is earlier (negative sign) or later (positive sign) than Coordinated Universal Time. Some nations (for example, Australia and India) have offset their time zones by half an hour. This situation is not shown on the illustration.*

Topographic Maps

A topographic map is a detailed and accurate two-dimensional representation of natural and human-made features on the Earth's surface. These maps are used for a number of applications, from camping, hunting, fishing, and hiking to urban planning, resource management, and surveying. The most distinctive characteristic of a topographic map is that the three-dimensional shape of the Earth's surface is modeled by the use of contour lines. Contours are imaginary lines that connect locations of similar elevation. Contours make it possible to represent the height of mountains and steepness of slopes on a two-dimensional map surface. Topographic maps also use a variety of symbols to describe both natural and human made features such as roads, buildings, quarries, lakes, streams, and vegetation.

Topographic maps produced by the Canadian National Topographic System (NTS) are generally available in two different scales: 1:50,000 and 1:250,000. Maps with a scale of 1:50,000 are relatively large-scale covering an area approximately 1000 square kilometers. At this scale, features as small as a single home can be shown. The smaller scale 1:250,000 topographic map is more of a general purpose reconnaissance-type map. A map of this scale covers the same area of land as sixteen 1:50,000 scale maps. In the United States, topographic maps have been made by the United States Geological

Survey (USGS) since 1879. Topographic coverage of the United States is available at scales of 1:24,000, 1:25,000 (metric), 1:62,250, 1:63,360 (Alaska only), 1:100,000 and 1:250,000.

Topographic Map Symbols

Topographic maps use symbols to represent natural and human constructed features found in the environment. The symbols used to represent features can be of three types: points, lines, and polygons. Points are used to depict features like bridges and buildings. Lines are used to graphically illustrate features that are linear. Some common linear features include roads, railways, and rivers. However, we also need to include representations of area, in the case of forested land or cleared land; this is done through the use of color. The set of symbols used on Canadian National Topographic System (NTS) maps has been standardized to simplify the map construction process. A description of the complete set of symbols available can be found in a published guide titled: Standards and Specifications for Polychrome Maps. This guide guarantees uniform illustration of surface features on both 1:50 000 and 1:250 000 topographic maps. Despite the existence of this guide, we can find that some topographic maps may use different symbols to depict a feature. This occurs because the symbols used are graphically refined over time - as a result the Standards and Specifications for Polychrome Maps guide is always under revision. The tables below describe some of the common symbols used on Canadian National Topographic System maps (source: Centre for Topographic Information, Natural Resources Canada). See the following link for the symbols commonly used on USGS topographic maps.

Transportation Features - Roads and Trails

Feature Name	Symbol
Road - hard surface, all season	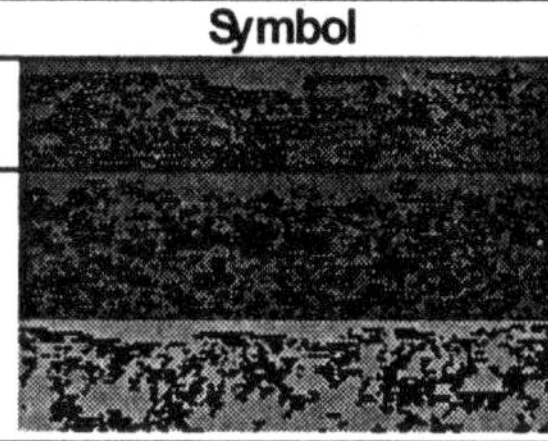
Road - hard surface, all season	
Road - loose or stabilized surface, all season	
Road - loose surface, dry weather	
Rapid transit route, road	
Road under construction	
Vehicle track or winter road	
Trail or portage	
Traffic circle	
Highway route number	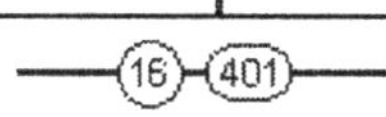

Transportation Features - Railways and Airports

Feature Name	Symbol
Railway - multiple track	
Railway - single track	
Railway sidings	
Railway - rapid transit	
Railway - under construction	
Railway - abandoned	
Railway on road	
Railway station	
Airfield; Heliport	
Airfield, position approximate	
Airfield runways; paved, unpaved	

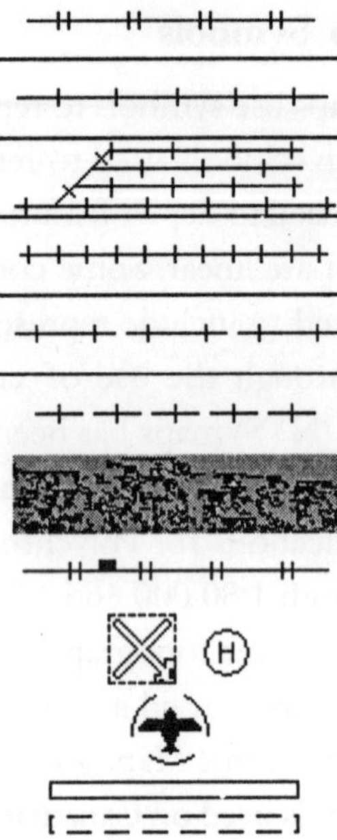

Other Transportation Features - Tunnels, Bridges, etc.

Feature Name	Symbol
Tunnel; railway, road	
Bridge	
Bridge; swing, draw, lift	
Footbridge	
Causeway	
Ford	
Cut	
Embankment	
Snow shed	
Barrier or gate	

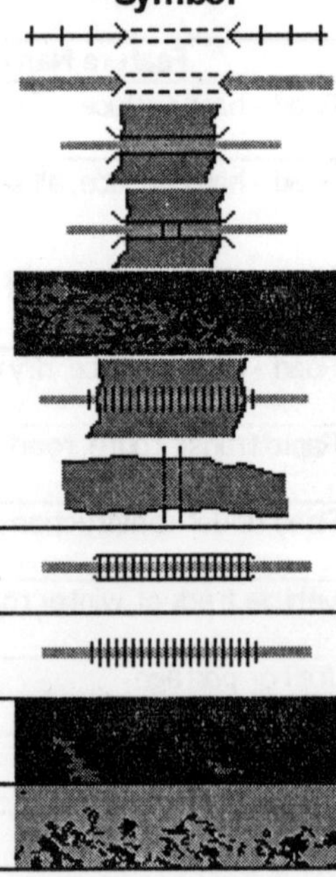

Hydrographic Features - Human Made

Feature Name	Symbol
Lock	
Dam; large, small	
Dam carrying road Footbridge Ferry Route	
Pier; Wharf; Seawall Breakwater Slip; Boat ramp; Drydock Canal; navigable or irrigation	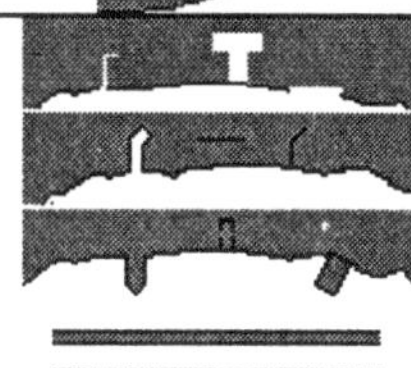
Canal, abandoned	
Shipwreck, exposed	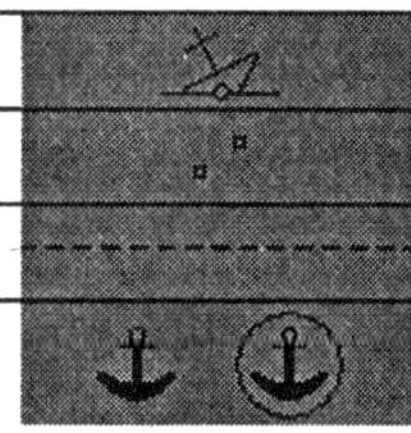
Crib or abandoned bridge pier	
Submarine cable	
Seaplane anchorage; Seaplane base	

Hydrographic Features - Naturally Occurring

Feature Name	Symbol
Falls	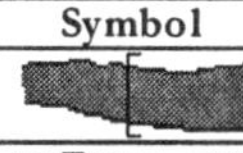
Rapids	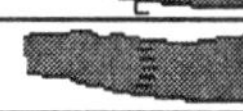
Direction of flow arrow Dry river bed Stream - intermittent Sand in Water or Foreshore Flats Rocky ledge, reef Flooded area Marsh, muskeg Swamp Well, water or brine; Spring Rocks in water or small islands Water elevation	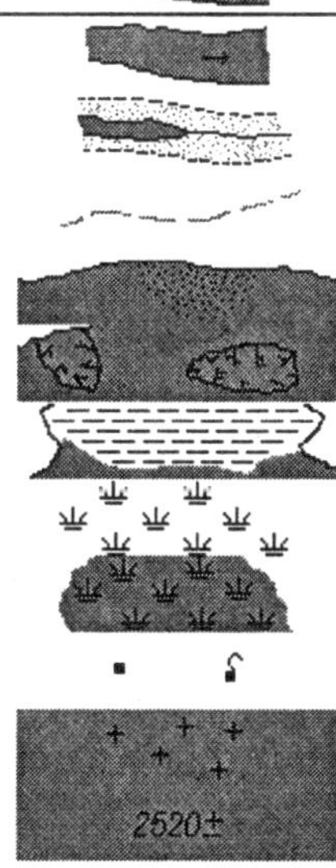

Terrain Features - Elevation

Feature Name	Symbol
Horizontal control point; Bench mark with elevation	△ 187.8↑
Precise elevation	▪224
Contours; index, intermediate	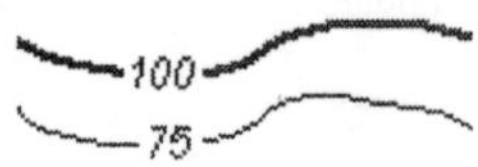
Depression contours	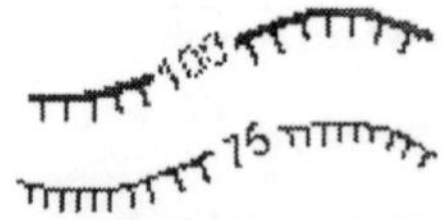

Terrain Features - Geology and Geomorphology

Feature Name	Symbol
Cliff or escarpment	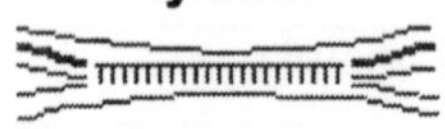
Esker	
Pingo	
Sand	
Moraine	
Quarry	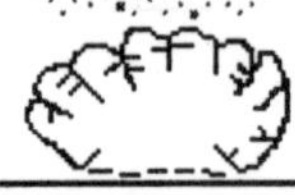
Cave	

Terrain Features - Land Cover

Feature Name	Symbol
Wooded area	
Orchard	
Vineyard	

CONTOUR LINES

Topographic maps can describe vertical information through the use of contour lines (contours). A contour line is an isoline that connects points on a map that have the same elevation. Contours are often drawn on a map at a uniform vertical distance. This distance is called the contour interval. The map in the Figure below shows contour lines with an interval of 100 feet. Note that every fifth brown contour lines is drawn bold and has the appropriate elevation labeled on it. These contours are called index contours. On Figure below they represent elevations of 500, 1000, 1500, 2000 feet and so on. The interval at which contours are drawn on a map depends on the amount of the relief depicted and the scale of the map.

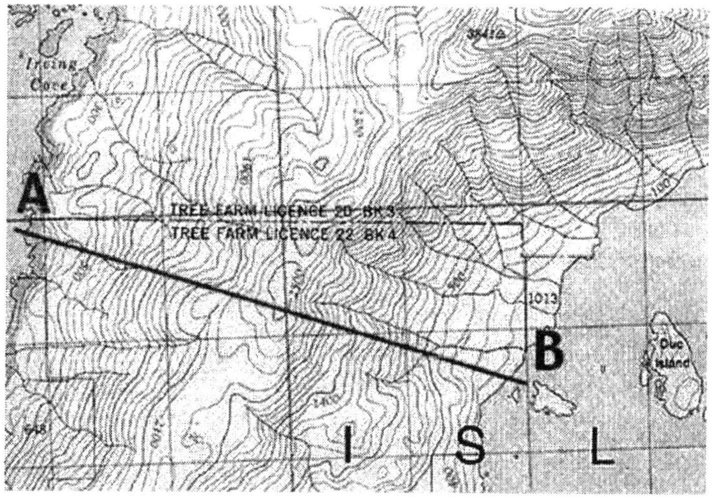

Figure: *Portion of the "Tofino" 1:50,000 National Topographic Series of Canada map. The brown lines drawn on this map are contour lines. Each line represents a vertical increase in elevation of 100 feet. The bold brown contour lines are called index contours. The index contours are labeled with their appropriate elevation which increases at a rate of 500 feet. Note the blue line drawn to separate water from land represents an elevation of 0 feet or sea-level.*

Contour lines provide us with a simple effective system for describing landscape configuration on a two-dimensional map. The arrangement, spacing, and shape of the contours provide the user of the map with some idea of what the actual topographic configuration of the land surface looks like. Contour intervals the are spaced closely together describe a steep slope. Gentle slopes are indicated by widely spaced contours. Contour lines that V upwards indicate the presence of a river valley. Ridges are shown by contours that V downwards.

Topographic Profiles

A topographic profile is a two-dimensional diagram that describes the landscape in vertical cross-section. Topographic profiles are often created from the contour information found on topographic maps. The simplest way to construct a topographic profile is to place a sheet of blank paper along a horizontal transect of interest. From the map, the elevation of the various contours is transferred on to the edge of the paper from one end of the transect to the other. Now on a sheet of graph paper use the x-axis to represent the horizontal distance covered by the transect. The y-axis is used to represent the vertical dimension and measures the change in elevation along the transect. Most people exaggerate the measure of elevation on the y-axis to make changes in relief stand out. Place the beginning of the transect as copied on the piece of paper at the intersect of the x and y-axis on the graph paper. The contour information on the paper's edge is now copied onto the piece of graph paper. Figure below shows a topographic profile drawn from the information found on the transect A-B above.

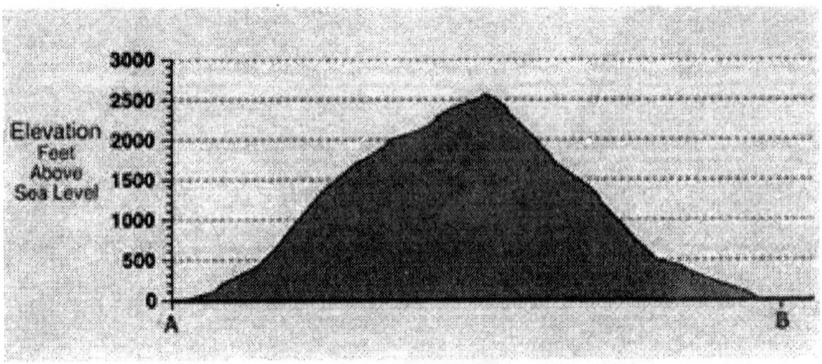

Figure: *The following topographic profile shows the vertical change in surface elevation along the transect AB from Figure 1d-1. A vertical exaggeration of about 4.2 times was used in the profile (horizontal scale = 1:50,000, vertical scale = 1:12,000 and vertical exaggeration = horizontal scale/vertical scale).*

MILITARY GRID REFERENCE SYSTEM AND MAP LOCATION

Two rectangular grid systems are available on topographic maps for identifying the location of points. These systems are the Universal Transverse Mercator (UTM) grid system and the Military Grid Reference System. The Military Grid Reference System is a simplified form of Universal Transverse Mercator grid system and it provides a very quick and easy method of referencing a location on a topographic map. On a topographic maps with a scale 1:50,000 and larger, the Military Grid Reference System is superimposed on the surface of map as blue colored series of equally spaced horizontal and vertical lines. Identifying numbers for each of these lines is found along the map's margin. Each identifying number consists of two digits which range from a value of 00 to 99. Each individual square in the grid system represents a distance of a 1000 by 1000 meters and the total size of the grid is 100,000 by 100,000 meters.

One problem associated with the Military Grid Reference System is the fact that reference numbers must be repeated every 100,000 meters. To overcome this difficulty, a method was devised to identify each 100,000 by 100,000 meter grid with two identifying letters which are printed in blue on the border of all topographic maps (note some maps may show more than one grid). When making reference to a location with the Military Grid Reference System identifying letters are always given before the horizontal and vertical coordinate numbers.

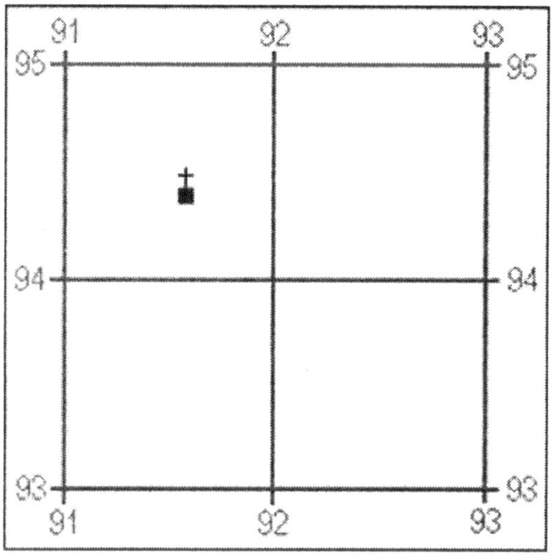

***Figure:** Portion of a Military Grid Reference System found on a topographic map. Coordinates on this system are based on a X (horizontal increasing from left to right) and Y (vertical increasing from bottom to top) system. The symbol depicting a church is located in the square 9194. Note that the value along the X-axis (easting) is given first followed by the value on the Y-axis (northing). (Source: Centre for Topographic Information, Natural Resources Canada).*

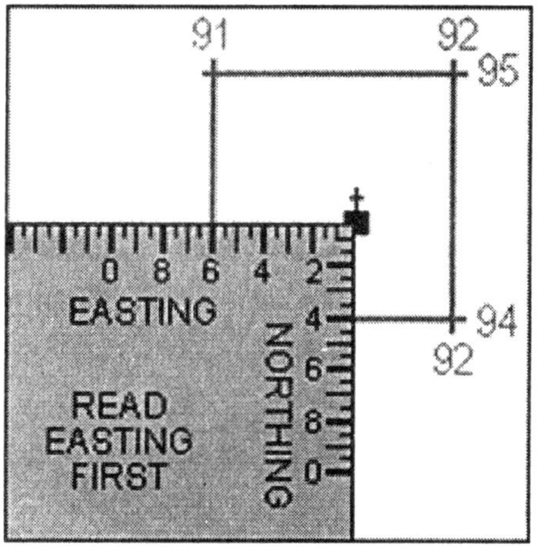

***Figure:** Further determination of the location of the church described in Figure above. Using the calibrated ruler we can now suggest the location of the church to be 916 on the X-axis and 944 on the Y-axis. Note that the location reference always has an even number of digits, with the three digits representing the easting and the second three the northing.*

Each individual square in the Military Grid Reference System can be further divided into 100 smaller squares (ten by ten). This division allows us to calculate the location of an object to within

100 meters. Figure below indicates that the church is six tenths of the way between lines 91 and 92, and four tenths of the way between lines 94 and 95. Using these values, we can state that the easting as being 916 and the northing as 944. By convention, these two numbers are combined into a coordinate reference of 916944.

INTRODUCTION TO REMOTE SENSING

Remote sensing can be defined as the collection of data about an object from a distance. Humans and many other types of animals accomplish this task with aid of eyes or by the sense of smell or hearing. Geographers use the technique of remote sensing to monitor or measure phenomena found in the Earth's lithosphere, biosphere, hydrosphere, and atmosphere. Remote sensing of the environment by geographers is usually done with the help of mechanical devices known as remote sensors.

These gadgets have a greatly improved ability to receive and record information about an object without any physical contact. Often, these sensors are positioned away from the object of interest by using helicopters, planes, and satellites. Most sensing devices record information about an object by measuring an object's transmission of electromagnetic energy from reflecting and radiating surfaces.

Table: Major regions of the electromagnetic spectrum.

Region Name	Wavelength	Comments
Gamma Ray	< 0.03 nanometers	Entirely absorbed by the Earth's atmosphere and not available for remote sensing.
X-ray	0.03 to 30 nanometers	Entirely absorbed by the Earth's atmosphere and not available for remote sensing.
Ultraviolet	0.03 to 0.4 micrometers	Wavelengths from 0.03 to 0.3 micrometers absorbed by ozone in the Earth's atmosphere.
Photographic Ultraviolet	0.3 to 0.4 micrometers	Available for remote sensing the Earth. Can be imaged with photographic film.
Visible	0.4 to 0.7 micrometers	Available for remote sensing the Earth. Can be imaged with photographic film.
Infrared	0.7 to 100 micrometers	Available for remote sensing the Earth. Can be imaged with photographic film.
Reflected Infrared	0.7 to 3.0 micrometers	Available for remote sensing the Earth. Near Infrared 0.7 to 0.9 micrometers. Can be imaged with photographic film.
Thermal Infrared	3.0 to 14 micrometers	Available for remote sensing the Earth. This wavelength cannot be captured with photographic film. Instead, mechanical sensors are used to image this wavelength band.
Microwave or Radar	0.1 to 100 centimeters	Longer wavelengths of this band can pass through clouds, fog, and rain. Images using this band can be made with sensors that actively emit microwaves.
Radio	> 100 centimeters	Not normally used for remote sensing the Earth.

Remote sensing imagery has many applications in mapping land-use and cover, agriculture, soils mapping, forestry, city planning, archaeological investigations, military observation, and geomorphological surveying, among other uses. For example, foresters use aerial photographs for preparing forest cover maps, locating possible access roads, and measuring quantities of trees harvested. Specialized photography using color infrared film has also been used to detect disease and insect damage in forest trees.

The simplest form of remote sensing uses photographic cameras to record information from visible or near infrared wavelengths. In the late 1800s, cameras were positioned above the Earth's surface in balloons or kites to take oblique aerial photographs of the landscape. During World War I, aerial photography played an important role in gathering information about the position and movements of enemy troops. These photographs were often taken from airplanes. After the war, civilian use of aerial photography from airplanes began with the systematic vertical imaging of large areas of Canada, the United States, and Europe. Many of these images were used to construct topographic and other types of reference maps of the natural and human-made features found on the Earth's surface. The development of color photography following World War II gave a more natural depiction of surface objects. Color aerial photography also greatly increased the amount of information gathered from an object. The human eye can differentiate many more shades of color than tones of gray. In 1942, Kodak developed color infrared film, which recorded wavelengths in the near-infrared part of the electromagnetic spectrum. This film type had good haze penetration and the ability to determine the type and health of vegetation.

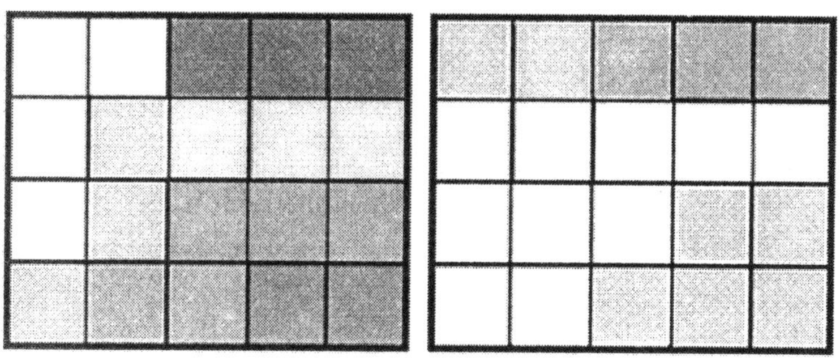

Figure: The rows of color tiles are replicated in the right as complementary gray tones. On the left, we can make out 18 to 20 different shades of color. On the right, only 7 shades of gray can be distinguished.

SATELLITE REMOTE SENSING

In the 1960s, a revolution in remote sensing technology began with the deployment of space satellites. From their high vantage-point, satellites have a greatly extended view of the Earth's surface. The first meteorological satellite, TIROS-1, was launched by the United States using an Atlas rocket on April 1, 1960. This early weather satellite used vidicon cameras to scan wide areas of the Earth's surface. Early satellite remote sensors did not use conventional film to produce their images. Instead, the sensors digitally capture the images using a device similar to a television camera. Once captured, this data is then transmitted electronically to receiving stations found on the Earth's surface. The image below is from TIROS-7 of a mid-latitude cyclone off the coast of New Zealand.

Today, the GOES (Geostationary Operational Environmental Satellite) system of satellites provides most of the remotely sensed weather information for North America. To cover the complete continent and adjacent oceans two satellites are employed in a geostationary orbit. The western half of North America and the eastern Pacific Ocean is monitored by GOES-10, which is directly above the equator and 135° West longitude. The eastern half of North America and the western Atlantic are cover by GOES-8. The GOES-8 satellite is located overhead of the equator and 75° West longitude. Advanced sensors aboard the GOES satellite produce a continuous data stream so images can be viewed at any instance. The imaging sensor produces visible and infrared images of the Earth's terrestrial surface and oceans. Infrared images can depict weather conditions even during the night. Another sensor aboard the satellite can determine vertical temperature profiles, vertical moisture profiles, total precipitable water, and atmospheric stability.

Figure: Color image from GOES-8 of hurricanes Madeline and Lester off the coast of Mexico, October 17, 1998. (Source: NASA - Looking at Earth From Space).

In the 1970s, the second revolution in remote sensing technology began with the deployment of the Landsat satellites. Since this 1972, several generations of Landsat satellites with their Multispectral Scanners (MSS) have been providing continuous coverage of the Earth for almost 30 years. Current, Landsat satellites orbit the Earth's surface at an altitude of approximately 700 kilometers. Spatial resolution of objects on the ground surface is 79 x 56 meters. Complete coverage of the globe requires 233 orbits and occurs every 16 days. The Multispectral Scanner records a zone of the Earth's surface that is 185 kilometers wide in four wavelength bands: band 4 at 0.5 to 0.6 micrometers, band 5 at 0.6 to 0.7 micrometers, band 6 at 0.7 to 0.8 micrometers, and band 7 at 0.8 to 1.1 micrometers. Bands 4 and 5 receive the green and red wavelengths in the visible light range of the electromagnetic spectrum. The last two bands image near-infrared wavelengths. A second sensing system was added to Landsat satellites launched after 1982. This imaging system, known as the Thematic Mapper, records seven wavelength bands from the visible to far-infrared portions of the electromagnetic spectrum. In addition, the ground resolution of this sensor was enhanced to 30 x 20 meters. This modification allows for greatly improved clarity of imaged objects.

The usefulness of satellites for remote sensing has resulted in several other organizations launching their own devices. In France, the SPOT (Satellite Pour l'Observation de la Terre) satellite program has launched five satellites since 1986. Since 1986, SPOT satellites have produced more than 10 million images. SPOT satellites use two different sensing systems to image the planet. One sensing system produces black and white panchromatic images from the visible band (0.51 to 0.73 micrometers) with a ground resolution of 10 x 10 meters. The other sensing device is multispectral capturing green, red, and reflected infrared bands at 20 x 20 meters. SPOT-5, which was launched in 2002, is much improved from the first four versions of SPOT satellites. SPOT-5 has a maximum ground resolution of 2.5 x 2.5 meters in both panchromatic mode and multispectral operation. Radarsat-1 was launched by the Canadian Space Agency in November, 1995. As a remote sensing device, Radarsat is quite different from the Landsat and SPOT satellites. Radarsat is an active remote sensing system that transmits and receives microwave radiation. Landsat and SPOT sensors passively measure reflected radiation at wavelengths roughly equivalent to those detected by our eyes. Radarsat's microwave energy penetrates clouds, rain, dust, or haze and produces images regardless of the Sun's illumination allowing it to image in darkness. Radarsat images have a resolution between 8 to 100 meters. This sensor has found important applications in crop monitoring, defence surveillance, disaster monitoring, geologic resource mapping, sea-ice mapping and monitoring, oil slick detection, and digital elevation modeling.

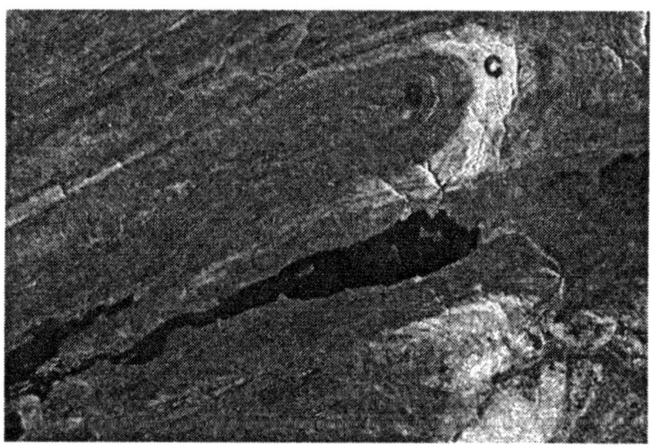

Figure: Radarsat image acquired on March 21, 1996, over Bathurst Island in Nunavut, Canada. This image shows Radarsat's ability to distinguish different types of bedrock. The light shades on this image (C) represent areas of limestone, while the darker regions (B) are composed of sedimentary siltstone. The very dark area marked A is Bracebridge Inlet which joins the Arctic ocean. (Source: Canadian Centre for Remote Sensing - Geological Mapping Bathurst Island, Nunavut, Canada March 21, 1996).

PRINCIPLES OF OBJECT IDENTIFICATION

Most people have no problem identifying objects from photographs taken from an oblique angle. Such views are natural to the human eye and are part of our everyday experience. However, most remotely sensed images are taken from an overhead or vertical perspective and from distances quite removed from ground level. Both of these circumstances make the interpretation of natural and human-made objects somewhat difficult. In addition, images obtained from devices that receive and capture electromagnetic wavelengths outside human vision can present views that are quite unfamiliar.

To overcome the potential difficulties involved in image recognition, professional image interpreters use a number of characteristics to help them identify remotely sensed objects. Some of these characteristics include:

Shape: this characteristic alone may serve to identify many objects. Examples include the long linear lines of highways, the intersecting runways of an airfield, the perfectly rectangular shape of buildings, or the recognizable shape of an outdoor baseball diamond.

Size: noting the relative and absolute sizes of objects is important in their identification. The scale of the image determines the absolute size of an object. As a result, it is very important to recognize the scale of the image to be analyzed.

Figure: Black and white aerial photograph of natural coniferous vegetation (left) and adjacent apple orchards (center and right).

Image Tone or Color: all objects reflect or emit specific signatures of electromagnetic radiation. In most cases, related types of objects emit or reflect similar wavelengths of radiation. Also, the types of recording device and recording media produce images that are reflective of their sensitivity to particular range of radiation. As a result, the interpreter must be aware of how the object being viewed will appear on the image examined. For example, on color infrared images vegetation has a color that ranges from pink to red rather than the usual tones of green.

Pattern: many objects arrange themselves in typical patterns. This is especially true of human-made phenomena. For example, orchards have a systematic arrangement imposed by a farmer, while natural vegetation usually has a random or chaotic pattern.

Shadow: shadows can sometimes be used to get a different view of an object. For example, an overhead photograph of a towering smokestack or a radio transmission tower normally presents an identification problem. This difficulty can be over come by photographing these objects at Sun angles that cast shadows. These shadows then display the shape of the object on the ground. Shadows can also be a problem to interpreters because they often conceal things found on the Earth's surface.

Texture: imaged objects display some degree of coarseness or smoothness. This characteristic can sometimes be useful in object interpretation. For example, we would normally expect to see textural differences when comparing an area of grass with a field corn. Texture, just like object size, is directly related to the scale of the image.

INTRODUCTION TO GEOGRAPHIC INFORMATION SYSTEMS

The advent of cheap and powerful computers over the last few decades has allowed for the development of innovative software applications for the storage, analysis, and display of geographic data. Many of these applications belong to a group of software known as Geographic Information Systems (GIS). Many definitions have been proposed for what constitutes a GIS. Each of these definitions conforms to the particular task that is being performed. Instead of repeating each of these definitions, I would like to broadly define GIS according to what it does. Thus, the activities normally carried out on a GIS include:

- The measurement of natural and human made phenomena and processes from a spatial perspective. These measurements emphasize three types of properties commonly associated with these types of systems: elements, attributes, and relationships.
- The storage of measurements in digital form in a computer database. These measurements are often linked to features on a digital map. The features can be of three types: points, lines, or areas (polygons).
- The analysis of collected measurements to produce more data and to discover new relationships by numerically manipulating and modeling different pieces of data.
- The depiction of the measured or analyzed data in some type of display - maps, graphs, lists, or summary statistics.

The first computerized GIS began its life in 1964 as a project of the Rehabilitation and Development Agency Program within the government of Canada. The Canada Geographic Information

System (CGIS) was designed to analyze Canada's national land inventory data to aid in the development of land for agriculture. The CGIS project was completed in 1971 and the software is still in use today. The CGIS project also involved a number of key innovations that have found their way into the feature set of many subsequent software developments.

From the mid-1960s to 1970s, developments in GIS were mainly occurring at government agencies and at universities. In 1964, Howard Fisher established the Harvard Lab for Computer Graphics where many of the industries early leaders studied. The Harvard Lab produced a number of mainframe GIS applications including: SYMAP (Synagraphic Mapping System),CALFORM, SYMVU, GRID, POLYVRT, and ODYSSEY. ODYSSEY was first modern vector GIS and many of its features would form the basis for future commercial applications.

Automatic Mapping System was developed by the United States Central Intelligence Agency (CIA) in the late 1960s. This project then spawned the CIA's World Data Bank, a collection of coastlines, rivers, and political boundaries, and the CAM software package that created maps at different scales from this data. This development was one of the first systematic map databases. In 1969, Jack Dangermond, who studied at the Harvard Lab for Computer Graphics, co-founded Environmental Systems Research Institute (ESRI) with his wife Laura. ESRI would become in a few years the dominate force in the GIS marketplace and create ArcInfo and ArcView software. The first conference dealing with GIS took place in 1970 and was organized by Roger Tomlinson (key individual in the development of CGIS) and Duane Marble (professor at Northwestern University and early GIS innovator). Today, numerous conferences dealing with GIS run every year attracting thousands of attendants.

In the 1980s and 1990s, many GIS applications underwent substantial evolution in terms of features and analysis power. Many of these packages were being refined by private companies who could see the future commercial potential of this software. Some of the popular commercial applications launched during this period include: ArcInfo, ArcView, MapInfo, SPANS GIS, PAMAP GIS, INTERGRAPH, and SMALLWORLD. It was also during this period that many GIS applications moved from expensive minicomputer workstations to personal computer hardware.

Components of a GIS

A Geographic Information System combines computer cartography with a database management system. Figure below describes some of the major components common to a GIS. This diagram suggests that a GIS consists of three subsystems: (1) an input system that allows for the collection of data to be used and analyzed for some purpose; (2) computer hardware and software systems that store the data, allow for data management and analysis, and can be used to display data manipulations on a computer monitor; (3) an output system that generates hard copy maps, images, and other types of output.

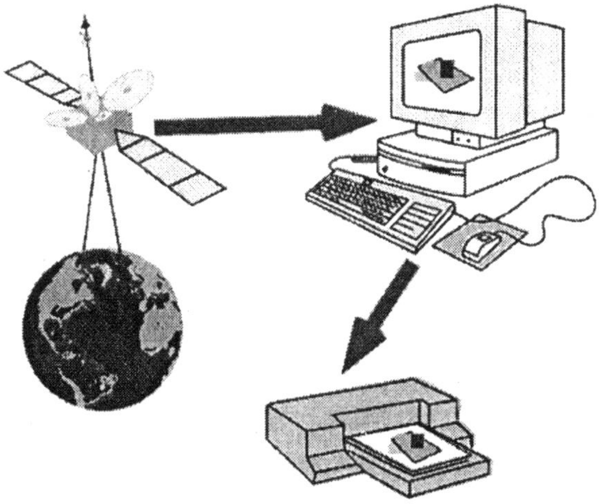

Figure: Three major components of a Geographic Information System. These components consist of input, computer hardware and software, and output subsystems.

Two basic types of data are normally entered into a GIS. The first type of data consists of real world phenomena and features that have some kind of spatial dimension. Usually, these data elements are depicted mathematically in the GIS as either points, lines, or polygons that are referenced geographically (or geocoded) to some type of coordinate system. This type data is entered into the GIS by devices like scanners, digitizers, GPS, air photos, and satellite imagery. The other type of data is sometimes referred to as an attribute. Attributes are pieces of data that are connected or related to the points, lines, or polygons mapped in the GIS. This attribute data can be analyzed to determine patterns of importance. Attribute data is entered directly into a database where it is associated with element data.

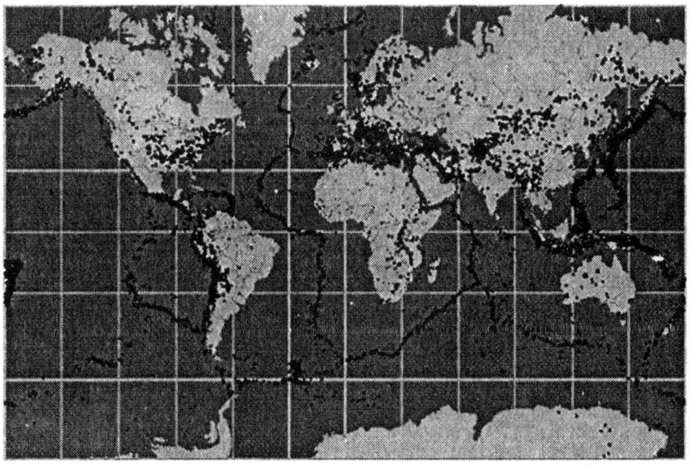

Figure: Distribution of earthquake events that have occurred over the last century.

These plotted data points can be defined as elements because their main purpose is to describe the location of the earthquakes.

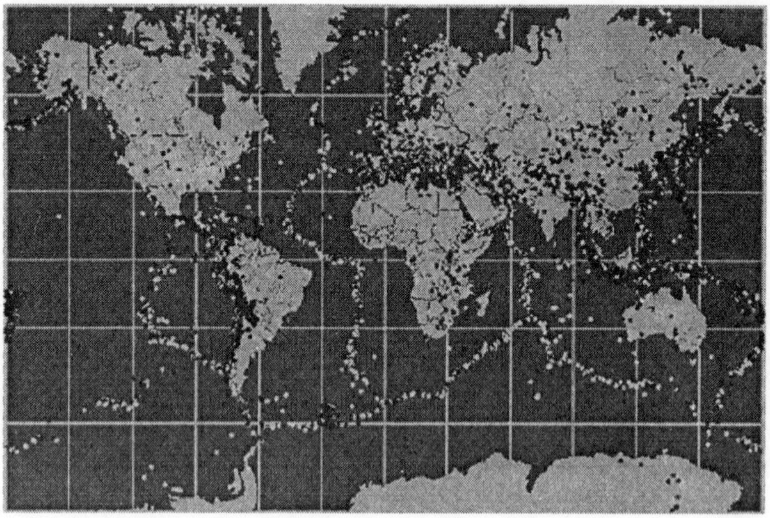

Figure: Earthquake events organized according to depth (yellow (shallow) = surface to 25 kilometers below the surface, red (intermediate) = 26 to 75 kilometers below the surface, and black (deep) = 76 to 660 kilometers below the surface).

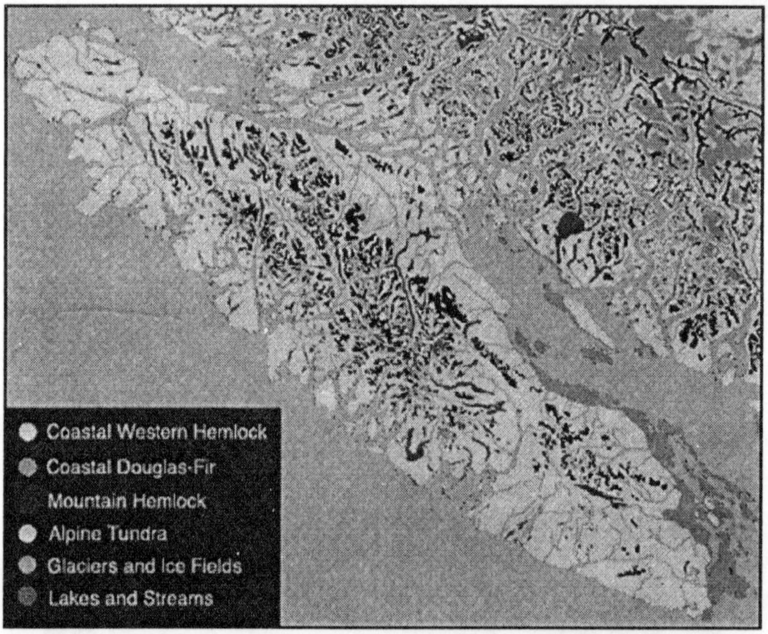

Figure: Graphic output from a GIS. This GIS contains information about the major plant communities, lakes and streams, and glaciers and ice fields found occupying the province of British Columbia, Canada. The output shows Vancouver Island and part of the British Columbia mainland.

For each of the earthquakes plotted on this map, the GIS also has data on their depth. These measurements can be defined as attribute data because they are connected to the plotted earthquake locations shows the attribute earthquake depth organized into three categories: shallow; intermediate; and deep. This analysis indicates a possible relationship between earthquake depth and spatial location - deep earthquakes do not occur at the mid-oceanic ridges.

Within the GIS database a user can enter, analyze, and manipulate data that is associated with some spatial element in the real world. The cartographic software of the GIS enables one to display the geographic information at any scale or projection and as a variety of layers which can be turned on or off. Each layer would show some different aspect of a place on the Earth. These layers could show things like a road network, topography, vegetation cover, streams and water bodies, or the distribution of annual precipitation received. The output illustrated in Figure below merges data layers for vegetation community type, glaciers and ice fields, and water bodies (streams, lakes, and ocean).

10

THE SCIENCE OF PHYSICAL GEOGRAPHY

SCIENTIFIC METHOD

Francis Bacon (1561-1626), a 17th century English philosopher, was the first individual to suggest a universal methodology for science. Bacon believed that scientific method required an inductive process of inquiry. Karl Popper later refuted this idea in the 20th century. Popper suggested that science could only be done using a deductive methodology. The next topic (3b) examines Karl Popper's recommended methodology for doing science more closely.

Science is simply a way of acquiring knowledge about nature and the Universe. To practice science, one must follow a specific universal methodology. The central theme of this methodology is the testing of hypotheses. A hypothesis can be defined as a proposal intended to explain certain facts or observations that has not been formally tested. The overall goal of science is to better comprehend the world around us. Various fields of study, like physics, chemistry, biology, medicine and the earth sciences, have used science exclusively to expand their knowledge base. Science allows its practitioners to acquire knowledge using techniques that are both neutral and unbiased.

The broadest, most inclusive goal of science is to understand (see Figure 3a-1). Understanding involves two interconnected processes: explanation and confirmation. Explanation is perhaps the most important basic goal of understanding.

Explanation consists of explaining reality with a system of hypotheses, theories, and laws. Explanation may also relate observed phenomena to a system of empirical formulas, or link them to mechanisms that are hierarchically structured at both higher and lower levels of function. A theory can be defined as a collection of logical ideas that are used to explain something. The process of testing, refining, and re-testing hypotheses constructs theories. The nature of this confirmation process suggests that theories are rarely static..

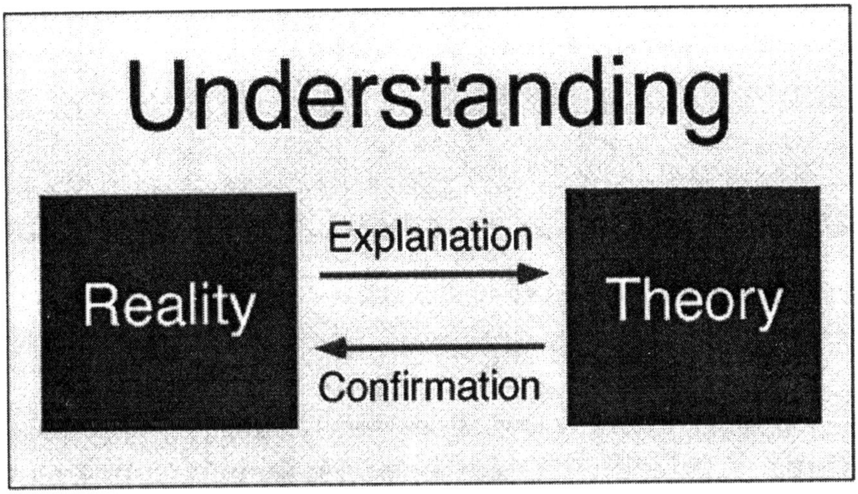

Figure: Relationship between reality, theory, and understanding in science. This model suggests that we develop scientific theories to explain phenomena found in reality. Once a theory is established, it must be confirmed by re-examining reality to find contrary data. If contrary data is found, the theory is modified to include this new information and the confirmation process begins again. The process of validating theories is endless process because we can never assume that we have considered all possibilities.

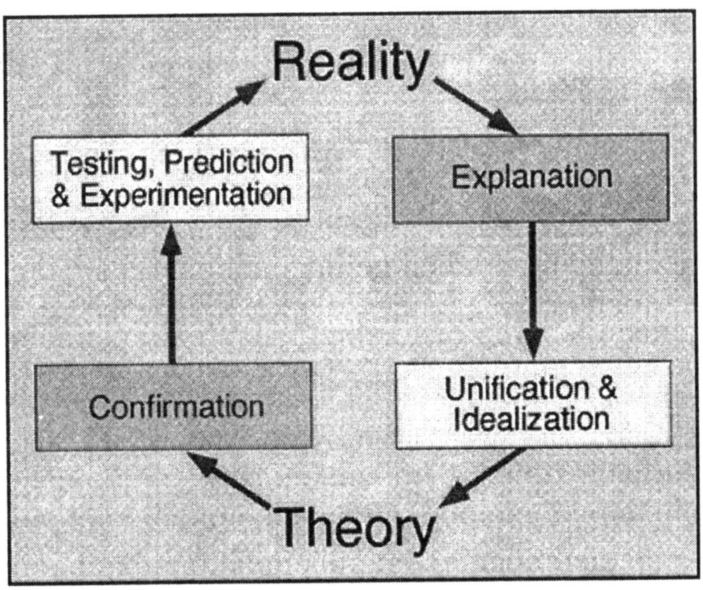

Figure: Facilitating tools involved explanation and confirmation.

Explanation has two important secondary components: idealization and unification. Idealization may be considered to be the condensation of a body of empirical fact into a simple statement. In the process of condensation, some detail must be omitted and the processes and phenomenon abstracted. Idealization may also involve isolating the phenomenon from other aspects of the system of interest. A second aspect of explanation is the unification of apparently unrelated phenomena in the same abstract or ideal system of concepts.

Another minor goal of science is the confirmation of constructed models or theories associated with understanding. Confirmation is accomplished through hypothesis testing, prediction, and by running experiments.

THE HYPOTHETICO-DEDUCTIVE METHOD

Philosopher Karl Popper suggested that it is impossible to prove a scientific theory true by means of induction, because no amount of evidence assures us that contrary evidence will not be found. Instead, Karl Popper proposed that proper science is accomplished by deduction. Deduction involves the process of falsification. Falsification is a particular specialized aspect of hypothesis testing. It involves stating some output from theory in specific and then finding contrary cases using experiments or observations. The methodology proposed by Popper is commonly known as the hypothetico-deductive method.

Popper's version of scientific method first begins with the postulation of a hypothesis. A hypothesis is an educated guess or a theory that explains some phenomenon. The researcher then tries to prove or test this scientific theory false through prediction or experimentation.A prediction is a forecast or extrapolation from the current state of the system of interest. Predictions are most useful if they can go beyond simple forecast. An experiment is a controlled investigation designed to evaluate the outcomes of causal manipulations on some system of interest.

To get a better understanding of the hypothetico-deductive method, we can examine the following geographic phenomena. In the brackish tidal marshes of the Pacific Coast of British Columbia and Washington, we find that the plants in these communities spatially arrange themselves in zones that are defined by elevation. Near the shoreline plant communities are dominated primarily by a single species known as Scirpus americanus. At higher elevations on the tidal marsh Scirpus americanus disappears and a species called Carex lyngbyei becomes widespread. The following hypothesis has been postulated to explain this unique phenomenon:

The distribution of Scirpus americanus and Carex lyngbyei is controlled by their tolerances to the frequency of tidal flooding. Scirpus americanus is more tolerant of tidal flooding than Carex lyngbyei and as a result it occupies lower elevations on the tidal marsh. However, Scirpus americanus cannot survive in the zone occupied by Carex lyngbyei because not enough flooding occurs. Likewise,

Carex lyngbyei is less tolerant of tidal flooding than Scirpus americanus and as a result it occupies higher elevations on the tidal marsh. Carex lyngbyei cannot survive in the zone occupied by Scirpus americanus because too much flooding occurs.

According to Popper, to test this theory a scientist would now have to prove it false. As discussed above this can be done in two general ways: (1) predictive analysis; or (2) by way of experimental manipulation. Each of these methods has been applied to this problem and the results are described below.

Predictive Analysis

If the theory is correct, we should find that in any tidal marsh plant community that contains Scirpus americanus and Carex lyngbyei that the spatial distribution of these two species should be similar in all cases. This is indeed true. However, there could be some other causal factor, besides flooding frequency, that may be responsible for these unique spatial patterns.

Experimental Manipulation

If the two species are transplanted into the zones of the other they should not be able to survive. An actual transplant experiment found that Scirpus americanus can actually grow in the zone occupied by Carex lyngbyei, while Carex lyngbyei could also grow at lower Scirpus sites. However, this growth became less vigorous as the elevation became lower and at a certain elevation it could not grow at all. These results falsify the postulated theory. So the theory must be modified based on the results and tested again.

The process of testing theories in science is endless. Part of this problem is related to the complexity of nature. Any one phenomenon in nature is influenced by numerous factors each having its particular cause and effect. For this reason, one positive test result is not conclusive proof that the phenomenon under study is explained. However, some tests are better than others and provide us with stronger confirmation. These tests usually allow for the isolation of the phenomena from the effects of causal factors. Manipulative experiments tend to be better than tests based on prediction in this respect.

Concepts of Time and Space in Physical Geography

The concepts of time and space are very important for understanding the function of phenomena in the natural world. Time is important to Physical Geographers because the spatial patterns they study can often only be explained in historic terms.

The measurement of time is not absolute. Time is perceived by humans in a relative fashion by using human created units of measurement. Examples of human created units of time are the measurement of seconds, minutes, hours, and days.

Geographers generally conceptualize two types of space. Concrete space represents the real world or environment. Abstract space models reality in a way that distills much of the spatial information contained in the real world. Maps are an excellent example of abstract space. Finally, like time, space is also perceived by humans in a relative fashion by using human created units of measurement.

Both time and space are variable in terms of scale. As such, researchers of natural phenomena must investigate their subjects in the appropriate temporal and/or spatial scales. For example, an investigator studying a forest ecosystem will have to deal with completely different scales of time and space when compared to a researcher examining soil bacteria. The trees that make up a forest generally occupy large tracts of land. For example, the boreal forest occupies millions of hectares in Northern Canada and Eurasia. Temporally, these trees have life spans that can be as long as several hundred years. On the other hand, soil bacteria occupy much smaller spatial areas and have life spans that can be measured in hours and days.

Study of Form or Process

Physical Geography as a science is experiencing a radical change in philosophy. It is changing from a science that was highly descriptive to one that is increasingly experimental and theoretical. This transition represents a strong desire by Physical Geographers to understand the processes that cause the patterns or forms we see in nature.

Before 1950, the main purpose of research in Physical Geography was the description of the natural phenomena. Much of this description involved measurement for the purpose of gaining basic facts dealing with form or spatial appearance. Out of this research Physical Geographers determined such things as: the climatic characteristics for specific locations and regions of the planet; flow rates of rivers; soil characteristics for various locations on the Earth's surface; distribution ranges of plant and animal species; and calculations of the amount of freshwater stored in lakes, glaciers, rivers and the atmosphere. By the beginning of the 20th century Physical Geographers began to examine the descriptive data that was collected, and started to ask questions related to why? Why is the climate of urban environments different from the climate of rural? Why does hail only form in thunderstorms? Why are soils of the world's tropical regions nutrient poor? Why do humid and arid regions of the world experience different levels of erosion?

In Physical Geography, and all other sciences, most questions that deal with why are usually queries about process. Some level of understanding about process can be derived from basic descriptive data. Process is best studied, however, through experimental manipulation and hypothesis testing. By 1950, Physical Geographers were more interested in figuring out process than just collecting descriptive facts about the world. This attitude is even more prevalent today because of our growing need to understand how humans are changing the Earth and its environment.

Finally, as mentioned above, a deeper understanding of process normally requires the use of hypothesis testing, experimental methods, and statistics. As a result, the standard undergraduate and graduate curriculum in Physical Geography exposes students to this type of knowledge so they can better ask the question why?

Descriptive Statistics

Physical Geographers often collect quantitative information about natural phenomena to further knowledge in their field of interest. This collected data is then often analyzed statistically to provide the researcher with impartial and enlightening presentation, summary, and interpretation of the phenomena understudy. The most common statistical analysis performed on data involves the determination of descriptive characteristics like measures of central tendency and dispersion.

It usually is difficult to obtain measurements of all the data available in a particular system of interest. For example, it may be important to determine the average atmospheric pressure found in the center of hurricanes. However, to make a definitive conclusion about a hurricane's central pressure with 100% confidence would require the measuring of all the hurricanes that ever existed on this planet. This type of measurement is called a population parameter. Under normal situations, the determination of population parameters is impossible, and we settle with a subset measure of the population commonly called an estimator. Estimators are determined by taking a representative sample of the population being studied.

Samples are normally taken at random. Random sampling implies that each measurement in the population has an equal chance of being selected as part of the sample. It also ensures that the occurrence of one measurement in a sample in no way influences the selection of another. Sampling methods are biased if the recording of some influences the recording of others or if some members of the population are more likely to be recorded than others.

11

THE SYSTEMS THEORY

HUMANS AND THEIR MODELS

One common conclusion of scientific inquiry is that the world of nature is often very complex. To understand this complexity, scientists usually try to envisage the phenomena of nature as simplified versions of reality known as a system. A system can be defined as a collection of interrelated parts that work together by way of some driving process.

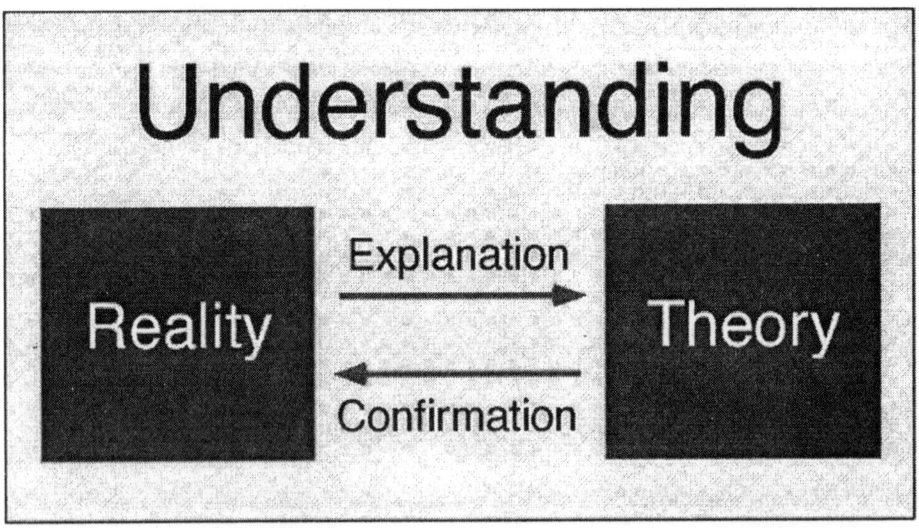

Figure 4a-1: The general relationship between perceptible phenomena and theory using scientific method for understanding. The interaction between perceptible phenomena and theory is arrived at through the processes of explanation and validation.

In the world of science, the word model is quite similar in meaning to the term system. Models in science tend to be simplified represenations of reality that can be explained ofmathematically and through the use of graphics. The following graphical model is used to help explain the processes involved in scientific understanding. The arrows in this graphically model suggest a continuous interaction between perceptible phenomena and theory through the processes of explanation and validation. This simple graphical model, while an extreme abstraction of the real world, is quite useful in explaining how scientific understanding works.

In Physical Geography, and many other fields of knowledge, systems and models are used extensively as aids in explaining natural phenomena around us.

DEFINITIONS OF SYSTEMS AND MODELS

As suggested in the previous section, a system is a assemblage of interrelated parts that work together by way of some driving process (see Figure 4b-1). Systems are often visualized or modeled as component blocks that have connections drawn between them. For example, the illustration below describes the interception of solar radiation by the Earth. In this system, the Earth and Sun, the parts or component blocks, are represented by two colored circles of different size. The process of solar emission and the interception of the Sun's emitted radiation by the Earth (the connection) is illustrated by the drawn lines.

Most systems share the same common characteristics. These common characteristics include the following:

1. Systems have a structure that is defined by its parts and processes.
2. Systems are generalizations of reality.
3. Systems tend to function in the same way. This involves the inputs and outputs of material (energy and/or matter) that is then processed causing it to change in some way.
4. The various parts of a system have functional as well as structural relationships between each other.
5. The fact that functional relationships exist between the parts suggests the flow and transfer of some type of energy and/or matter.
6. Systems often exchange energy and/or matter beyond their defined boundary with the outside environment, and other systems, through various input and output processes.
7. Functional relationships can only occur because of the presence of a driving force.
8. The parts that make up a system show some degree of integration - in other words the parts work well together.

Within the boundary of a system we can find three kinds of properties:

Elements - are the kinds of parts (things or substances) that make up a system. These parts may be atoms or molecules, or larger bodies of matter like sand grains, rain drops, plants, animals, etc.

Attributes - are characteristics of the elements that may be perceived and measured. For example: quantity, size, color, volume, temperature, and mass.

Relationships - are the associations that occur between elements and attributes. These associations are based on cause and effect.

We can define the state of the system by determining the value of its properties (the elements, attributes, and/or relationships).

Scientists have examined and classified many types of systems. Some of the classified types include:

Isolated System - a system that has no interactions beyond its boundary layer. Many controlled laboratory experiments are this type of system.

Closed System - is a system that transfers energy, but not matter, across its boundary to the surrounding environment. Our planet is often viewed as a closed system.

Open System - is a system that transfers both matter and energy can cross its boundary to the surrounding environment. Most ecosystems are example of open systems.

Morphological System - this is a system where we understand the relationships between elements and their attributes in a vague sense based only on measured features or correlations. In other words, we understand the form or morphology a system has based on the connections between its elements. We do not understand exactly how the processes work to transfer energy and/or matter through the connections between the elements.

Cascading System - this is a system where we are primarily interested in the flow of energy and/or matter from one element to another and understand the processes that cause this movement. In a cascading system, we do not fully understand quantitative relationships that exist between elements related to the transfer of energy and/or matter.

Process-Response System - this is a system that integrates the characteristics of both morphological and cascading systems. In a process-response system, we can model the processes involved in the movement, storage, and transformation of energy and/or matter between system elements and we fully understand how the form of the system in terms of measured features and correlations.

Control System - a system that can be intelligently manipulated by the action of humans.

Ecosystem - is a system that models relationships and interactions between the various biotic and abiotic components making up a community or organisms and their surroundng physical environment.

Structure of Systems

Systems exist at every scale of size and are often arranged in some kind of hierarchical fashion. Large systems are often composed of one or more smaller systems working within its various elements.

Processes within these smaller systems can often be connected directly or indirectly to processes found in the larger system. A good example of a system within systems is the hierarchy of systems found in our Universe. Let us examine this system from top to bottom:

At the highest level in this hierarchy we have the system that we call the cosmos or Universe. The elements of this system consist of galaxies, quasars, black holes, stars, planets and other heavenly bodies. The current structure of this system is thought to have come about because of a massive explosion known as the Big Bang and is controlled by gravity, weak and strong atomic forces, and electromagnetic forces.

Around some stars in the universe we have an obvious arrangement of planets, asteroids, comets and other material. We call these systems solar systems. The elements of this system behave according to set laws of nature and are often found orbiting around a central star because of gravitational attraction. On some planets conditions may exist for the development of dynamic interactions between the hydrosphere, lithosphere, atmosphere, or biosphere.

We can define a planetary system as a celestial body in space that orbits a star and that maintains some level of dynamics between its lithosphere, atmosphere and hydrosphere. Some planetary systems, like the Earth, can also have a biosphere. If a planetary system contains a biosphere, dynamic interactions will develop between this system and the lithosphere, atmosphere and hydrosphere. These interactions can be called an environmental system. Environmental systems can also exist at smaller scales of size (e.g., a single flower growing in a field could be an example of a small-scale environmental system).

The Earth's biosphere is made up small interacting entities called ecosystems. In an ecosystem, populations of species group together into communities and interact with each other and the abiotic environment. The smallest lining entity in an ecosystem is a single organism. An organism is alive and functioning because it is a biological system.

The elements of a biological system consist of cells and larger structures known as organs that work together to produce life. The functioning of cells in any biological system is dependent on numerous chemical reactions.

Together these chemical reactions make up a chemical system. The types of chemical interactions found in chemical systems are dependent on the atomic structure of the reacting matter. The components of atomic structure can be described as an atomic system.

ENVIRONMENTAL SYSTEMS AS ENERGY SYSTEMS

In the previous section we define an environmental system as a system where life interacts with the various abiotic components found in the atmosphere, hydrosphere, and lithosphere. Environmental systems also involve the capture, movement, storage, and use of energy. Thus, environmental systems are also energy systems.

In environmental systems, energy moves from the abiotic environment to life through processes like plant photosynthesis. Photosynthesis packages this energy into simple organic compounds like glucose and starch. Both of these organic molecules can be stored for future use. The following chemical formula describes how plants capture the Sun's light energy and convert it into chemical energy:

$$6CO_2 + 6H_2O + \text{light energy} = C_6H_{12}O_6 + 6O_2$$

The energy of light is used by plants in this reaction to chemically change carbon dioxide (CO_2) and water (H_2O) into oxygen (O_2) and the energy rich organic molecule glucose ($C_6H_{12}O_6$).

The chemical energy of photosynthesis can be passed on to other living or biotic components of an environmental system through biomass consumption or decomposition by consumer organisms. When needed for metabolic processes, the fixed organic energy stored in an organism can be released to do work via respiration or fermentation.

Energy also fuels a number of environmental processes that are essentially abiotic. For example, the movement of air by wind, the weathering of rock into soil, the formation of precipitation, and the creation of mountains by tectonic forces.

The first three processes derive their energy directly or indirectly from the Sun's radiation that is received at the Earth's surface. Mountain building is fueled by the heat energy that exists within the Earth's interior.

Finally, the movement of energy in environmental systems always obeys specific thermodynamic laws that cannot be broken. We will learn more about these laws later in this textbook.

Food Chain as an Example of a System

A food chain models the movement of energy in an ecosystem (a form of environmental system). Figure 4e-1 below illustrates the movement of energy in a typical food chain. In this diagram, we begin the food chain with 100,000 units of light energy from the Sun. Note, the amount of energy available at each successive level (called trophic levels) of this system becomes progressively less. Only 10 units of energy are available at the last level (carnivores) of the food chain. A number of factors limit the assimilation of energy from one level to the next. We will examine these factors later in this textbook.

The Sun is the original source of energy, in the form of light, for the food chain. (100,000 Units of Energy)

Plants capture approximately 1% of the available light energy from the Sun for biomass production by way of photosynthesis. Photosynthesis can be described chemically as: Light Energy + $6CO_2$ + $6H_2O$ ==> $C_6H_{12}O_6$ + $6O_2$ (1,000 Units of Energy)

Herbivores consume approximately 10% of the plant biomass produced in a typical food chain.(100 Units of Energy)

Carnivores capture and consume about 10% of the energy stored by the herbivores.(10 Units of Energy)

Figure: Model of the grazing food chain showing the movement of energy through an ecosystem.

Why is the above illustration an example of a system? The concept of what makes something a system was fully explained in previous section. In this topic, it was suggested that all systems share the following seven common characteristics:

1. Systems have a structure that is defined by its parts and processes. In the above example, the structure consists of the system's three types of properties. This system has the following elements: the Sun, plants, herbivores and carnivores. Within this system the main characteristic, or attribute, of the elements being perceived is units of energy. The last component that makes

up the structure of this system is the cause and effect relationships between the elements and attributes. For example, the Sun creates energy via nuclear fusion. This energy is radiated from the Sun's surface and received by the surface of the Earth. On the surface of the Earth plants capture some of this solar radiation in the chloroplasts that exist in their tissues. Through photosynthesis the plants convert the radiant energy into energy rich organic matter. Some of the energy fixed by the plants is passed on to herbivores through consumption. Finally, a portion of the energy assimilated by the herbivores is then passed on to carnivores through consumption.

2. Systems are generalizations of reality. The food chain process described above is a simple abstraction of what actually happens in a variety of different types of terrestrial ecosystems of much greater complexity.
3. Systems tend to function in the same way. All systems consist of groups of parts that interact with each other according to various cause and effect processes. In the food chain model, the parts are the Sun, plants, herbivores and carnivores. There are two main processes taking place in this system. The first involves the movement of energy, in the form of radiation, from the Sun to the plants. The second process involves the movement of energy, in the form of organic molecules, from plants to herbivores, and then finally to carnivores through biomass consumption.
4. The various parts of a system have functional as well as structural relationships between each other. The structure within the food chain is defined by the functional relationships between the elements and attributes of the system.
5. The fact that functional relationships exist between the parts suggests the flow and transfer of some type of energy or matter. Systems exchange energy and matter internally and with their surrounding environment through various processes of input and output. The main material being transferred into this system (input) is energy in the form of solar radiation. The solar radiation is then fixed into organic matter (output) by way of photosynthesis in the plants. Herbivores consume the constructed plant organic molecules for nutrition to run their metabolism. The herbivores then provide food for the carnivores.
6. Systems often exchange energy and/or matter beyond their defined boundary with the outside environment, and other systems, through various input and output processes. The organisms found in a food chains transfer organic matter into the detritus food chain when they shed tissues or die. This transfer represnts a net output of matter out of the food chain. With decomposition, the organic matter is converted into inorganic nutrients which can be taken up by plants in the food chain to produce new organic matter. This transfer represents a net input of matter into the food chain system.

7. Functional relationships can only occur because of the presence of a driving force. The driving force in the food chain is the Sun.
8. The parts that make up a system show some degree of integration - in other words the parts work well together. Integration in the food chain comes primarily from the process of evolution. It was through evolution that plants, herbivores, and carnivores came about and developed ecological associations between each other.

EQUILIBRIUM CONCEPTS AND FEEDBACKS

Equilibrium

Equilibrium describes the average condition of a system, as measured through one of its elements or attributes, over a specific period of time. For the purposes of this online textbook, there are six types of equilibrium:

(1) Steady state equilibrium is an average condition of a system where the trajectory remains unchanged in time.

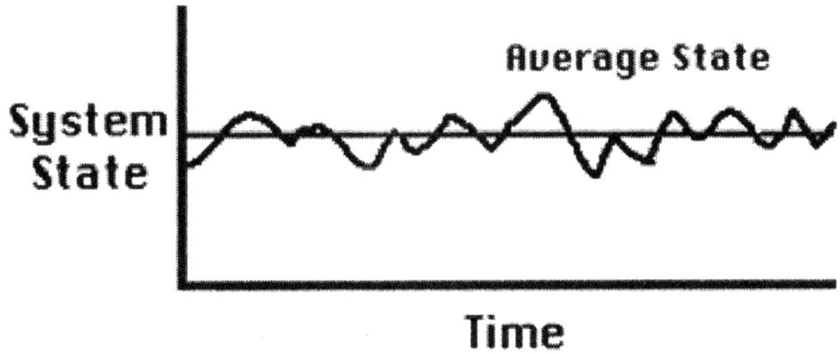

Figure: Example of the state of a steady state equilibrium over time.

(2) Thermodynamic equilibrium describes a condition in a system where the distribution of mass and energy moves towards maximum entropy.

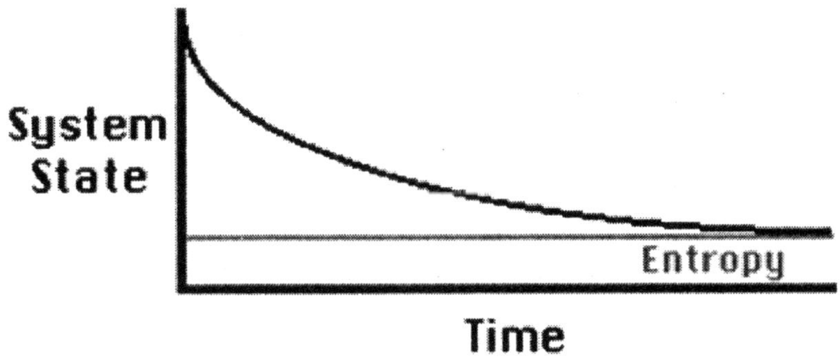

Figure: Example of the state of a thermodynamic equilibrium over time.

(3) A dynamic equilibrium occurs when there are unrepeated average states through time.

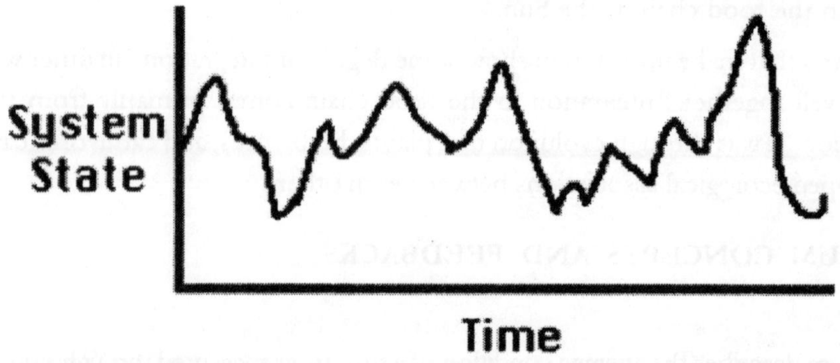

Figure: Example of the state of a dynamic equilibrium over time.

(4) Static equilibrium occurs where force and reaction are balanced and the properties of the system remain unchanged over time.

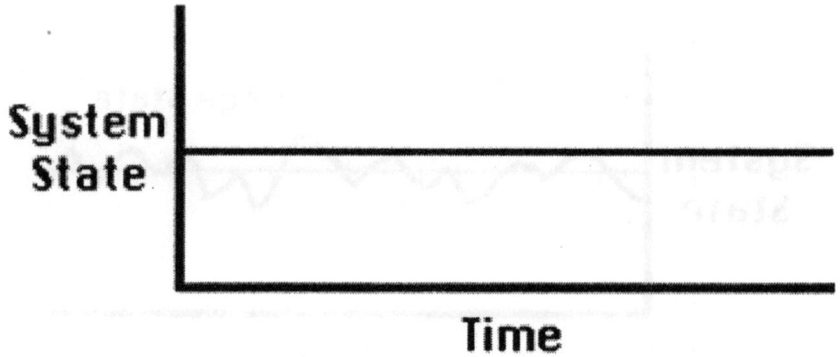

Figure: Example of the state of a static equilibrium over time.

(5) In a stable equilibrium the system displays tendencies to return to the same equilibrium after disturbance.

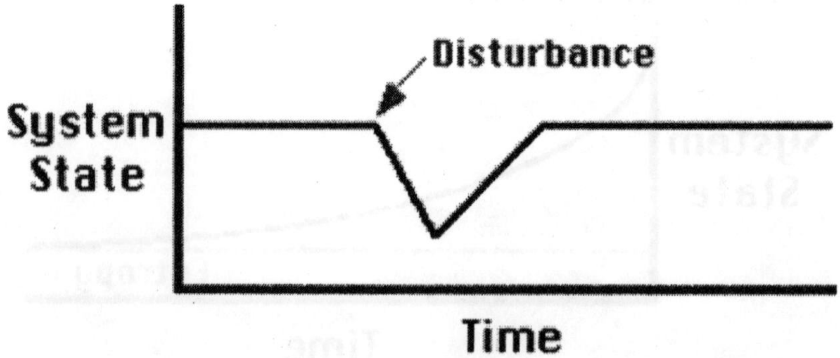

Figure: Example of the state of a stable equilibrium over time.

(6) In an unstable equilibrium the system returns to a new equilibrium after disturbance.

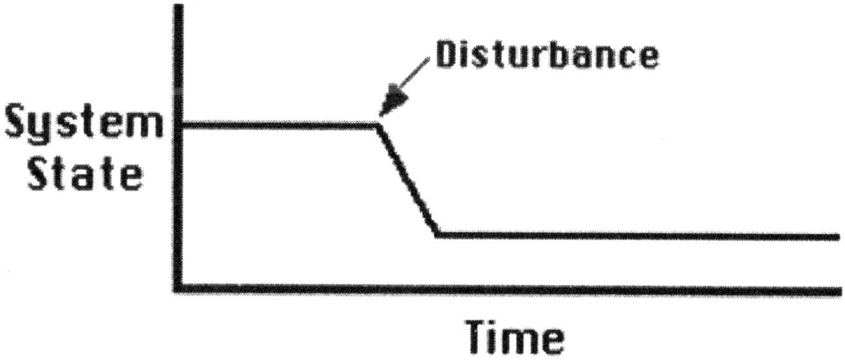

Figure: Example of the state of an unstable equilibrium over time.

Feedbacks

In order for a system to maintain a steady state or average condition the system must possess the capacity for self-regulation. Self-regulation in many systems is controlled by negative feedback and positive feedback mechanisms. Negative-feedback mechanisms control the state of the system by dampening or reducing the size of the system's elements or attributes. Positive-feedback mechanisms feed or increase the size of one or more of the system's elements or attributes over time.

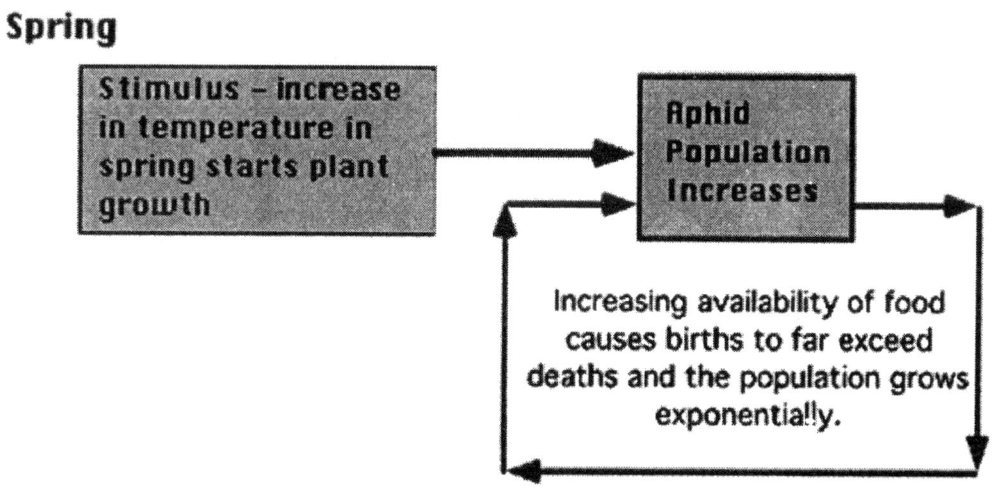

Figure: Increase in food abundance causes a positive effect on the size of the aphid population. Births increase significantly and are much higher than deaths causing the population to expand.

Interactions among living organisms, or between organisms and the abiotic environment typically involve both positive feedback and negative feedback responses. Feedback occurs when an organism's system state depends not only on some original stimulus but also on the results of its previous system state. Feedback can also involve the system state of non-living components in an ecosystem. A positive feedback causes a self-sustained change that increases the state of a system. Negative feedback causes the system to decease its state over time. The presence of both negative and positive feedback mechanisms in a system results in self-regulation.

To illustrate how these mechanisms work we can hypothetically examine the changes in aphid population growth in a mid-latitude climate in the following graphical models.

Over the winter only a few aphids survive the cold winter temperatures and scarcity of food. However, warmer temperatures in the spring cause plants to start growing, providing the aphids with food.

The continually increasing abundance of food increases the fertility of individual aphids and the aphid population starts to expand exponentially. This situation is a positive feedback as abundant food resources cause increased reproduction and rapid population growth.

By late summer the supply of plants has reached its maximum. As a result, the availability of food to the growing aphid population becomes less per individual as time proceeds. Less food means lower egg production and births begin to decline. When births reach the same level as deaths the aphid population stops growing and stabilizes. The population is becoming too large for the food supply. The size of the population relative to food supply produces a negative feedback as the fertility of aphids begins to decline.

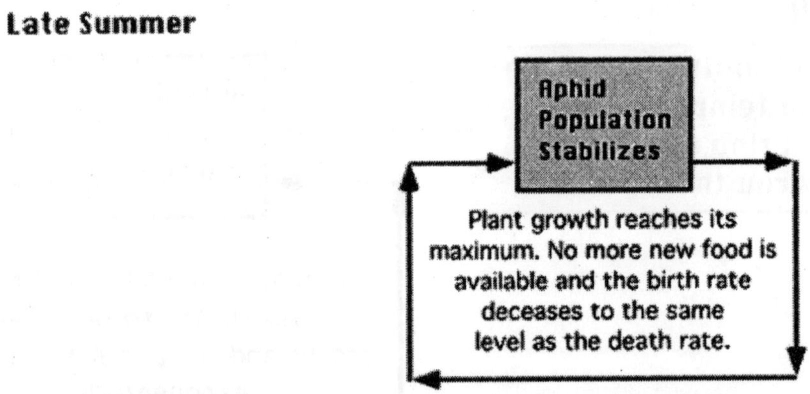

Figure: As the abundance of food levels off aphid reproduction slows down and deaths begin to increase. Population size of the aphids begins to level off.

With the arrival of fall cooler temperatures reduce the supply of plants causing the large aphid population to be faced with further food scarcity. The lack of food causes the aphid death rate to increase and the birth rate to decrease. The population begins to decline rapidly. The reduction of food caused by the cooler temperatures enhances the negative feedback that began in late summer.

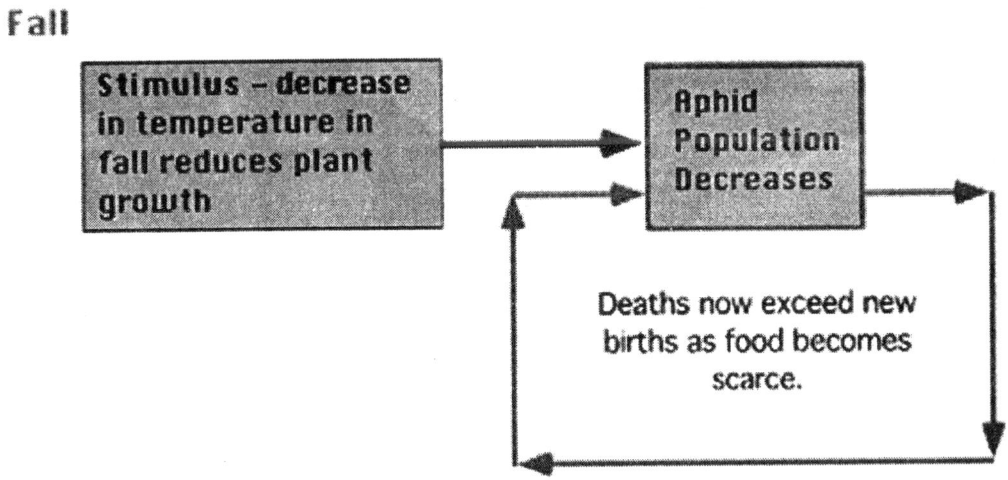

Figure: The decrease in food abundance causes a negative effect on aphid population size. Deaths now exceed births and the population size of the aphids begins to decrease.

12

THE UNIVERSE, EARTH, NATURAL SPHERES, AND GAIA

EVOLUTION OF THE UNIVERSE

About 11 to 15 billion years ago all of the matter and energy in the Universe was concentrated into an area the size of an atom. At this moment, matter, energy, space and time did not exist. Then suddenly, the Universe began to expand at an incredible rate and matter, energy, space and time came into being (the Big Bang). As the Universe expanded, matter began to coalesce into gas clouds, and then stars and planets. Our solar system formed about 5 billion years ago when the Universe was about 65% of its present size. Today, the Universe continues to expand.

WHY DO MOST SCIENTISTS ACCEPT THE BIG BANG THEORY?

The acceptance of this theory by the scientific community is based on a number of observations. These observations confirm specific predictions of the Big Bang theory. In a previous section, we learned that scientists test their theories through deduction and falsification. Predictions associated with the Big Bang theory that have been tested by this process are:

1. If the Big Bang did occur, all of the objects within the Universe should be moving away from each other. In 1929, Edwin Hubble documented that the galaxies in our Universe are indeed moving away from each other.
2. The Big Bang should have left an "afterglow" from the explosion. In the 1960s, scientists discovered the existence of cosmic background radiation, the so-called "afterglow" after the Big Bang explosion. Our most accurate measurements of this cosmic radiation came in November 1989, by the Cosmic Background Explorer (COBE) satellite. The measurements

from this satellite tested an important prediction of the Big Bang theory. This prediction suggests that the initial explosion that gave birth to the Universe should have created radiation with a spectrum that follows a blackbody curve. The COBE measurements indicated that the spectrum of the cosmic radiation varied from a blackbody curve by only 1%. This level of error is considered insignificant.

3. If the Universe began with a Big Bang, extreme temperatures should have caused 25 percent of the mass of the Universe to become helium. This is exactly what is observed.
4. Matter in the Universe should be distributed homogeneously. Astronomical observations from the Hubble Space Telescope do indicate that matter in the Universe generally has a homogeneous distribution.

How will the Universe End?

Cosmologists have postulated two endings to the Universe. If the Universe is infinite or has no edge, it should continue to expand forever. A Universe that is finite or closed is theorized to collapse when expansion stops because of gravity. The collapse of the Universe ends when all matter and energy is compressed into the high energy, high-density state from which it began. This scenario is of course called the Big Crunch. Some theorists have suggested that the Big Crunch will produce a new Big Bang and the process of an expanding Universe will begin again. This idea is called the oscillating Universe theory.

Early History of the Earth

Scientists believe the Earth began its life about 4.6 billion years ago. The Earth formed as cosmic dust lumped together to form larger and larger particles until 150 million years had passed. At about 4.4 billion years, the young Earth had a mass similar to the mass it has today. The continents probably began forming about 4.2 billion years ago as the Earth continued to cool.

The cooling also resulted in the release of gases from the lithosphere, much of which formed the Earth's early atmosphere. Most of the Earth's early atmosphere was created in the first one million years after solidification (4.4 billion years ago). Carbon dioxide, nitrogen, and water vapor dominated this early atmosphere. Table below describes the three major stages of development of the atmosphere.

As the Earth continued to cool, the water vapor found in the atmosphere condensed to form the oceans and other fresh water bodies on the continents. Oxygen began accumulating in the atmosphere through photo-dissociation of O_2 from water, and by way of photosynthesis (life). The emergence of living organisms was extremely important in the creation of atmospheric oxygen and ozone. Without ozone, life could not exist on land because of harmful ultraviolet radiation.

Table: Evolution of the Earth's atmosphere.

Name of Stage	Duration of Stage (Billions of Years Ago)	Main Constituents of the Atmosphere	Dominant Processes and Features
Early Atmosphere	4.4 to 4.0	H_2O, hydrogen cyanide (HCN), ammonia (NH_3), methane (CH_4), sulfur, iodine, bromine, chlorine, argon	Lighter gases like hydrogen and helium escaped to space. All water was held in the atmosphere as vapor because of high temperatures.
Secondary Atmosphere	4.0 to 3.3	At 4.0 billion H_2O, CO_2, and nitrogen (N) dominant. Cooling of the atmosphere causes precipitation and the development of the oceans. By 3.0 billion CO_2, H_2O, N_2 dominant. O_2 begins to accumulate.	Continued release of gases from the lithosphere. Water vapor clouds common in the lower atmosphere. Chemosynthetic bacteria appear on the Earth some time between 3.9 and 3.5 billion years ago. Life begins to modify the atmosphere.
Living Atmosphere	3.3 to Present	N_2 - 78%, O_2 - 21%, Argon - 0.9%, CO_2 - 0.036%	Development, evolution and growth of life increases the quantity of oxygen in the atmosphere from <1% to 21%. 500 million years ago concentration of atmospheric oxygen levels off. Humans begin modifying the concentrations of some gases in the atmosphere beginning around the year 1700.

Most of the build up of oxygen in the atmosphere occurred between 2.1 and 1.5 billion years ago as a direct result of photosynthesis from ocean based plants like algae.

At about 450 million years ago, there was enough oxygen in the atmosphere to allow for the development of a stratospheric ozone layer that was thick enough to keep terrestrial life protected from ultraviolet radiation.

As a result, terrestrial life began its development and expansion at this time. Table 5b-2 describes the timing of the evolutionary development of some of the Earth's dominant forms of life before and after 450 million years before present (BP).

Table: Approximate origin time of the major plant and animal groups.

Organism Group	Time of Origin
Marine Invertebrates	570 Million Years Ago
Fish	505 Million Years Ago
Land Plants	438 Million Years Ago
Amphibians	408 Million Years Ago
Reptiles	320 Million Years Ago
Mammals	208 Million Years Ago
Flowering Plants (Angiosperms)	140 Million Years Ago

The Natural Spheres

From the standpoint of Physical Geography, the Earth can be seen to be composed of four principal components:

- Lithosphere - describes the solid inorganic portion of the Earth (composed of rocks, minerals and elements). It can be regarded as the outer surface and interior of the solid Earth. On the surface of the Earth, the lithosphere is composed of three main types of rocks:
  - Igneous - rocks formed by solidification of molten magma.
  - Sedimentary - rocks formed by the alteration and compression of old rock debris or organic sediments.
  - Metamorphic - rocks formed by alteration of existing rocks by intense heat or pressure.
- Atmosphere - is the vast gaseous envelope of air that surrounds the Earth. Its boundaries are not easily defined. The atmosphere contains a complex system of gases and suspended particles that behave in many ways like fluids. Many of its constituents are derived from the Earth by way of chemical and biochemical reactions.

- Hydrosphere - describes the waters of the Earth (see the hydrologic cycle). Water exists on the Earth in various stores, including the atmosphere, oceans, lakes, rivers, soils, glaciers, and groundwater. Water moves from one store to another by way of: evaporation, condensation, runoff, precipitation, infiltration and groundwater flow.
- Biosphere - consists of all living things, plant and animal. This zone is characterized by life in profusion, diversity, and ingenious complexity. Cycling of matter in this sphere involves not only metabolic reactions in organisms, but also many abiotic chemical reactions.

All of these spheres are interrelated to each other by dynamic interactions, like biogeochemical cycling, that move and exchange both matter and energy between the four components.

The Gaia Hypothesis

In 1965, J.E. Lovelock published the first scientific paper suggesting the Gaia hypothesis. The Gaia hypothesis states that the temperature and composition of the Earth's surface are actively controlled by life on the planet.

It suggests that if changes in the gas composition, temperature or oxidation state of the Earth are caused by extraterrestial, biological, geological, or other disturbances, life responds to these changes by modifying the abiotic environment through growth and metabolism.

In simplier terms, biological responses tend to regulate the state of the Earth's environment in their favor.

The evidence for Gaia is as follows:

- If not continually replaced by biotic activities gases like methane and hydrogen would become non-existant in the atmosphere in a few decades.
- Carbon dioxide (CO_2) in the Earth's atmosphere is far less abundant than chemistry alone would allow. If life was deleted carbon dioxide would become 30 times more abundant. Large quantities of carbon dioxide are currently locked up by living organisms.
- The Sun's energy output has increased by 30% in the past 3.5 billion years. Yet, historical climate data indicates that the temperature of the Earth has only fluctuated by about 5° Celsius from the current average global temperature of 15° Celsius. Computer climate models suggest that a 30% reduction in solar radiation would create a global average temperature of between -10 and -52° Celsius all things being equal. These results indicate that levels of atmospheric carbon dioxide must have been much higher in the past when the Sun was less powerful. Extra atmospheric carbon dioxide would have created a greater greenhouse effect and warmer temperatures. These results also indicate that some mechanism must have removed carbon dioxide from the atmosphere as the Sun's output of radiation increased over the

Earth's geologic history. This mechanism is the conversion of atmospheric carbon dioxide into fossilized organic matter (natural gas, oil, coal, limestone, and peat). In other words, Gaia!

This theory is important to Physical Geography and other Earth Sciences for the following reasons:

- The Gaia theory suggests that the abiotic and biotic environment is made up of many complex interrelationships;
- Many of these complex interrelationships are quite delicate and may be altered by human activity to a breaking point; and
- The theory suggests that humans must learn to respect Gaia by reducing their intentional modification of the Earth's abiotic and biotic components.

13

ENERGY AND MATTER

CHARACTERISTICS OF ENERGY AND MATTER

Energy is defined simply by scientists as the capacity for doing work. Matter is the material (atoms and molecules) that constructs things on the Earth and in the Universe. Albert Einstein suggested early in this century that energy and matter are related to each other at the atomic level. Einstein theorized that it should be possible to convert matter into energy. From Einstein's theories, scientists were able to harness the energy of matter beginning in the 1940s through nuclear fission. The most spectacular example of this process is a nuclear explosion from an atomic bomb. A more peaceful example of our use of this fact of nature is the production of electricity from controlled fission reactions in nuclear reactors. Einstein also suggested that it should be possible to transform energy into matter.

Energy and matter are also associated to each other at much larger scales of nature. Later on in this chapter, we will examine how solar radiation provides the energy to create the matter that makes up organisms. Organisms then use some of this matter to power their metabolism.

Types of Energy

Energy comes in a variety of forms. The simplest definition of the types of energy suggests that two forms exist: kinetic energy and potential energy. Kinetic energy is the energy due to motion. A rock falling from a cliff, a bee in flight, wind blowing leaves of trees, and water following over a waterfall are all examples of kinetic energy. Potential energy is the energy stored by an object that can be potentially transformed into another form of energy. Water stored behind a dam, the chemical energy of the food we consume, and the gasoline that we putting in our cars are all examples of potential energy. Conversion of this energy occurs when the energy in food is used by an organism to energize its metabolism, when the water in the dam flows through turbines to produce electricity from motion, and when the gasoline is used in a engine to produce motion from combustion.

Some other forms of energy include heat, electricity, sound, energy of chemical reactions, magnetic attraction, energy of atomic reactions, and light. Definitions for a few of these types of energy are as follows:

- Radiation - is the emission of energy from a material object in the form of electromagnetic waves and photons.
- Chemical Energy - is the energy produced or consumed in chemical reactions.
- Atomic Energy - is the energy released from an atomic nucleus because of a change in its subatomic mass.
- Electrical Energy - is the energy produced from the force between two objects having the physical property of electrical charge.
- Heat Energy - is a form of energy created by the combined internal motion of atoms in a substance.

On Earth, there are fundamentally only three ways in which energy can be transferred from one place to another: conduction, convection, and radiation. Conduction involves the adjacent transfer of heat energy from one atom to another through the mass of a gas, liquid, or solid. Condution results in the continuous flow of heat energy along a temperature gradient from areas of higher to lower temperature. Convection involves the transfer of heat energy by way of mass movements of a substance in gas or liquid form in a vertical direction (horizontal transfer is called advection). Convection is often seen as rising masses of gas or liquid called convection currents. It is important to note that energy transfer by way of conduction and convection depends on the presence of matter. These forms of energy transfer do not operate in the vacuum of space. Radiation is the only means of energy transfer that can occur across outer space. The transfer of radiation produced at the Sun's surface through space supplies the Earth with most of its energy.

MATTER: ELEMENTS AND COMPOUNDS

Matter is the material that makes up things in the Universe. All matter on the Earth is constructed of elements. Chemists have described approximately 115 different elements. Each of these elements have distinct chemical characteristics. Table below lists some of the chemical characteristics for 48 common elements found in the Earth's continental crust.

The smallest particle that exhibits the unique chemical characteristics of an element is known as an atom. Atoms are composed of yet smaller particles known as protons, neutrons, and electrons. A proton is a subatomic particle that has significant mass and contributes a single positive electrical charge to an atom. Neutrons also have significant mass but no electrical charge. Electrons are extremely light subatomic particles having a mass that is 1/1840 of a proton. Each electron also has a negative electrical charge.

Table: Characteristics of some of the common chemical elements found in the Earth's continental crust.

Element	Chemical Symbol	Atomic Number	Common Atomic Mass Number	Atomic Weight	Percent in Continental Crust	Required for all Life	Required for Some Life forms	Element Type	Moderately Toxic	Extremely Toxic
Aluminum	Al	13	27	26.98	8.2300	-	X	Metalloid	-	-
Antimony	Sb	51	122	121.75	0.00002	-	-	Metalloid	-	-
Arsenic	As	33	75	74.92	0.00018	-	-	Metalloid	-	X
Barium	Ba	56	137	137.34	0.0425	-	-	Metal	-	-
Beryllium	Be	4	10	9.01	0.00028	-	-	Metal	-	X
Bismuth	Bi	83	209	208.98	0.000017	-	-	Metal	-	-
Boron	B	5	11	10.81	0.0010	-	-	Metalloid	-	-
Bromine	Br	35	80	79.91	0.00025	-	-	Nonmetal	-	-
Cadmium	Cd	48	112	112.40	0.00002	-	-	Metal	-	X
Calcium	Ca	20	40	40.08	4.1000	X	-	Metal	-	-
Carbon	C	6	12	12.01	0.0200	X	-	Nonmetal	-	-
Chlorine	Cl	17	35.5	35.45	0.0130	-	X	Nonmetal	X	-
Chromium	Cr	24	52	52.00	0.0100	-	-	Metal	X	-
Cobalt	Co	27	59	58.93	0.0025	-	X	Metal	-	-
Copper	Cu	29	63.5	63.54	0.0055	X	-	Metal	X	-
Fluorine	F	9	19	19.00	0.0625	-	X	Nonmetal	X	-
Gallium	Ga	31	70	69.72	0.0015	-	-	Metal	-	-
Germanium	Ge	32	73	72.59	0.00015	-	-	Metalloid	-	-
Gold	Au	79	197	196.97	0.0000004	-	-	Metal	-	-
Hydrogen	H	1	1	1.008	1.4000	X	-	Nonmetal	-	-
Iodine	I	53	127	126.90	0.00005	-	X	Nonmetal	-	-
Iron	Fe	26	56	55.85	5.6000	X	-	Metal	-	-
Lead	Pb	82	207	207.19	0.00125	-	-	Metal	-	X
Lithium	Li	3	6	6.94	0.0020	-	-	Metal	-	-
Magnesium	Mg	12	24	24.31	2.3000	X	-	Metal	-	-
Manganese	Mn	25	55	54.94	0.0950	X	-	Metal	-	-
Mercury	Hg	80	201	200.59	0.000008	-	-	Metal	-	X
Molybdenum	Mo	42	96	95.94	0.00015	X	-	Metal	-	-
Nickel	Ni	28	59	58.71	0.0075	-	-	Metal	-	X
Nitrogen	N	7	14	14.01	0.0020	X	-	Nonmetal	-	-
Oxygen	O	8	16	16.00	46.4000	X	-	Nonmetal	-	-
Palladium	Pd	46	106	106.40	0.000001	-	-	Metal	X	-
Phosphorus	P	15	31	30.97	0.1050	X	-	Nonmetal	-	-
Platinum	Pt	78	195	195.09	0.0000005	-	-	Metal	-	-
Potassium	K	19	39	39.10	2.1000	X	-	Metal	-	-
Rubidium	Rb	37	85.5	85.47	0.0090	-	-	Metal	-	-
Selenium	Se	34	79	78.96	0.000005	-	X	Nonmetal	X	-
Silicon	Si	14	28	28.09	28.2000	-	-	Metalloid	-	-
Silver	Ag	47	108	107.87	0.000007	-	-	Metal	-	X
Sodium	Na	11	23	22.99	2.4000	-	X	Metal	-	-
Sulfur	S	16	32	32.06	0.0260	X	-	Nonmetal	-	-
Thorium	Th	90	232	232.04	0.00096	-	-	-	-	-
Tin	Sn	50	119	118.69	0.00020	-	-	Metal	X	-
Titanium	Ti	22	48	47.90	0.5700	-	-	Metal	-	-
Tungsten	W	74	184	183.85	0.00015	-	-	Metal	-	-
Uranium	U	92	238	238.03	0.00027	-	-	-	-	-
Vanadium	V	23	51	50.94	0.0135	-	X	Metal	X	-
Zinc	Zn	30	65	65.37	0.0070	X	-	Metal	-	-

Protons and neutrons make up the nucleus of an atom. As a result, most of an atom's mass is concentrated in the nucleus. Because protons are positively charged the nucleus has a positive charge equal to the number of these subatomic particles.

Electrons are found orbiting outside the nucleus at various distances based on their energy level. The area occupied by the electrons has a negative charge equal to the number of these subatomic particles.

If an atom has an equal number of electrons and protons its net electrical charge is zero. If there are more electrons than protons the charge of the atom is negative. Likewise, if there are less electrons than protons the charge of the atom is positive. In both cases, the exact charge is determined by subtracting protons from electrons. As a result, 4 protons minus 6 electrons give an atomic charge of -2.

The number of protons found in the nuclei of the different types of elements is unique and is referred to as the atomic number.

All atoms of a specific element have the same number of protons in their nuclei. Atomic mass number is an atom's total number of neutrons and protons. Many elements have unequal numbers of neutrons and protons in their nucleus. An element's atomic weight refers to the total weight of neutrons, protons, and electrons.

For example, the atomic weight of aluminum is 26.98. Atomic number describes the number of protons found in an atom. For example, silver has an atomic number of 47 or 47 protons in its atom. Some elements can have variants containing different numbers of neutrons but similar numbers of protons.

We call these variants isotopes. Carbon has two isotopes. Its most common form is carbon-12 which has 6 protons plus 6 neutrons.

About 99% of the carbon on our planet is of this type. The isotope carbon-13 has 6 protons plus 7 neutrons. Carbon-14 is the rarest isotope of carbon containing 8 neutrons. Some isotopes are unstable and their nucleus tends to lose subatomic particles forming an element with a lower atomic mass. This process is known as radioactive decay.

Elements can be classified as being either metals, nonmetals, or metalloids. Metals are elements that usually conduct heat and electricity and are shiny. Nonmetals do not conduct electricity that well and are normally not shiny. Metalloids have characteristics that are in between metals and nonmetals.

Elements with a net positive or negative charge are called ions. Chemists indicate the number of positive or negative charges on an ion using a superscript after the element's symbol. For example, calcium has two positive charges and is written as Ca^{2+}. Some common negatively charged ions include nitrate (NO^{3-}), sulfate (SO_4^{2-}), and phosphate (PO_4^{3-}).

Positive and negative ions are electrically attracted to each other. This mutual attraction allows for the bonding of atoms to occur forming structures of matter that are larger than just one atom. When similar atoms bond together they construct molecules. Atoms of different elements joined together form compounds. Sodium chloride (or table salt), is an ionic compound consisting of sodium (Na^+) and chloride (Cl^-). In nature, it forms as a three-dimensional array of oppositely charged ions. Many of the Earth's substances have a molecular structure similar to sodium chloride.

Inorganic vs. Organic

Compounds and molecules constructed in living tissues are commonly called organic. Forms of matter not formed by living things are termed inorganic. Organisms like autotrophs usually create organic matter by consuming inorganic molecules and compounds from the lithosphere, hydrosphere, and atmosphere. An example of an autotroph is any photosynthesizing plant. Heterotrophs consume and assimilate other living things to create their organic matter. Herbivores and carnivores are examples of heterotrophs.

There are four general categories of organic compounds: lipids, carbohydrates, proteins, and nucleic acids.

- Lipids - are composed of carbon atoms that have two hydrogen atoms attached. Lipids are commonly known as fats and oils, and belong to the family of molecules known as hydrocarbons.
- Carbohydrates - are composed of carbon, oxygen, and hydrogen atoms. Some examples are sugars, starch, and cellulose.
- Proteins - are organic compounds that are made primarily of carbon, hydrogen, nitrogen, and some other minor elements that are arranged into 20 different compounds known as amino acids.
- Nucleic Acids - are composed primarily of different combinations of carbon, hydrogen, nitrogen, oxygen, and phosphorus. They are very complex compounds being created by the atomic linking of thousands of individual atoms. DNA or deoxyribonucleic acid, the genetic blueprint of life, is an example of a nucleic acid.

Cells

All organisms are composed of one or more of cells. Cells are the smallest self-functioning unit found in living organisms. Cells are also where the processes of metabolism and heredity occur in an organism. Cells arise by the cellular division of a previously existing cell. Biologists have differentiated two basic types of cells in organisms. Bacteria, archaea, and cyanobacteria have cells that are quite uncomplicated in terms of structure and function. Quite simply, they lack internal organization. These cells are commonly known as prokaryotes.

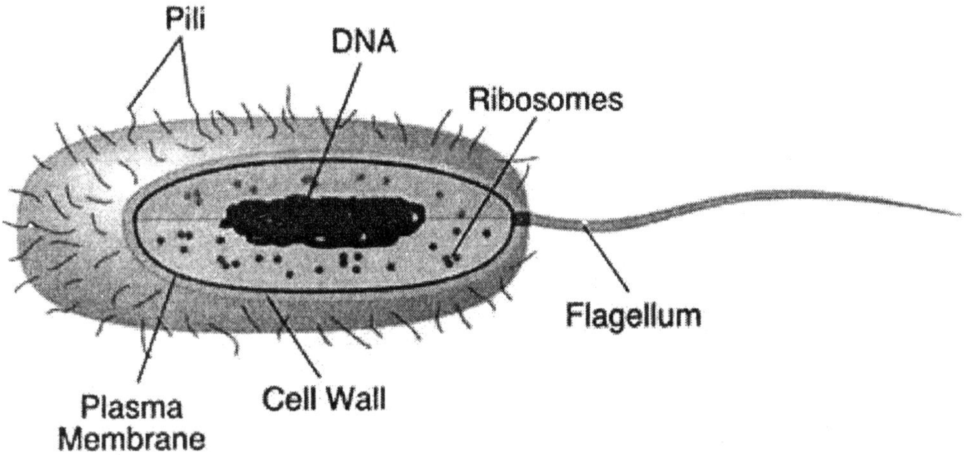

Figure: Typical prokaryote cell. These cells are about 1 to 10 micrometers in size. The cell is encased by rigid cell wall and a plasma membrane. Within the cell, the two most obvious structures are ribosomes and DNA. The DNA is not bounded by a membrane. Many prokaryote cells also have a flagellum which is used for movement.

The cells of plants and animals are more complex than those of bacteria, archaea, and cyanobacteria. We call these type of cells eukaryotes. Eukaryotic cells have a membrane enclosed nucleus which contains the organism's DNA. Plant and animal cells also contain a variety of membrane-bound structures known as organelles.

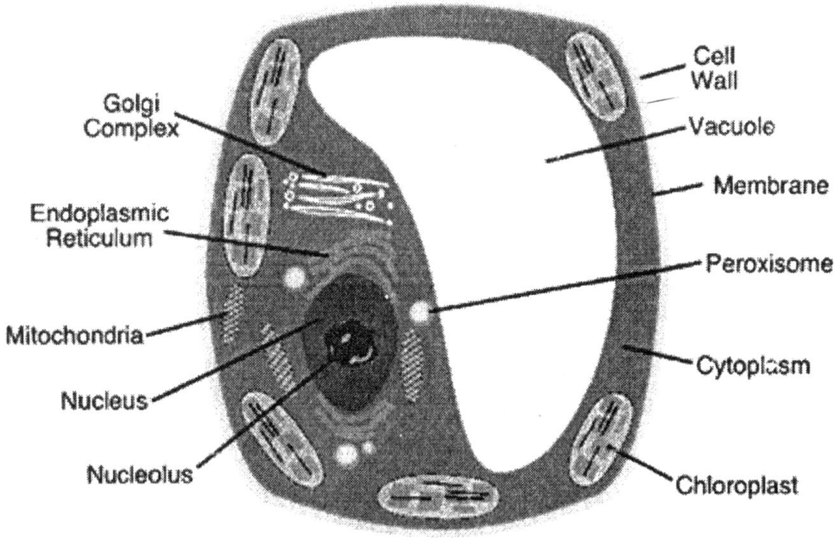

Figure: Typical plant cell. Plants cells differ from animal cells in the following ways: they have a cell wall, chloroplasts, and they often contain a large central vacuole.

Eukaryotic cells also show great variation in terms of size. White blood cells of mammals are some of the smallest eukaryotic cells with a diameter between 3 to 4 micrometers. Ostrich ovum are very large cells with a diameter of about 100 micrometers.

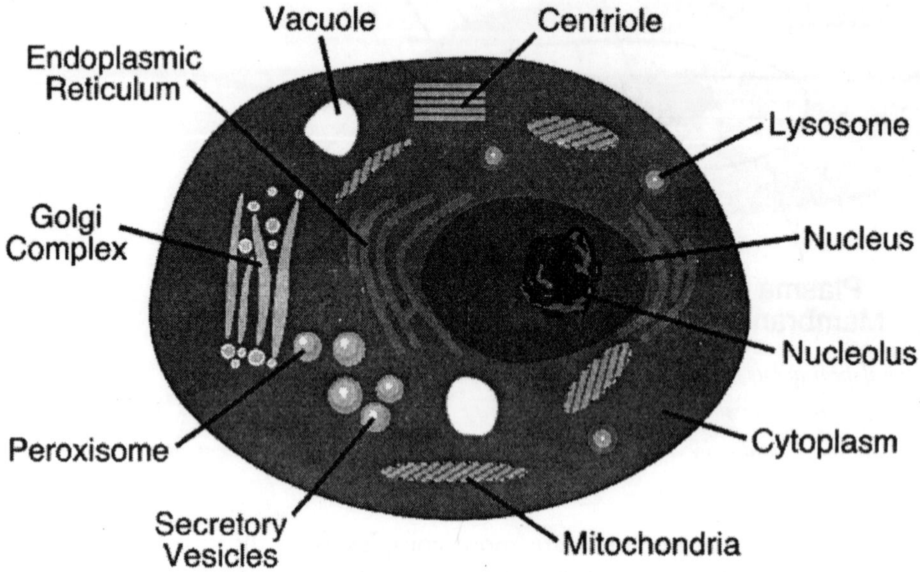

Figure: Typical animal cell.

Internally, cells contain specialized structures known as organelles that carry out distinct cellular functions. Within these structures enzymes, a type of protein, are used to facilitate and regulate various chemical reactions. Table below describes the function of a variety of cell structures including many organelles.

Cells can also be classified according to how they obtain their energy. Some cells have the ability to use light or chemical energy found in the outside environment to manufacture their own sugars, fats, and proteins. We call these types of cells autotrophs.

All 400,000 species of plants and a few species of bacteria use sunlight and the process of photosynthesis to obtain their energy. Some bacteria breakdown molecules found in the environment to release chemical energy to sustain their life. Organisms can also obtain their energy by consuming other organisms. These organisms are called heterotrophs. Heterotrophs include most types of bacteria and all of the animal and fungi species.

Some organisms consist of just one cell (bacteria, algae, and protozoa). However, most organisms are multi-cellular. Within multi-cellular organisms, groups of cells can become specialized to carry out a certain function. We call these functional groups of cells a tissue. An organ is a structure composed of several different types of tissues. Organs also have a specific structure and a particular function.

Table: Description and function of common cell structures.

Structure	Description	Function
Cell Wall	Outer layer on a cell composed of cellulose or other complex carbohydrates.	Helps to support and protect the cell.
Plasma Membrane	A layer composed of lipids and proteins that controls the permeability of the cell to water and dissolved substances.	Regulates the movement of material into and out of the cell.
Flagellum (Flagella pl.)	Threadlike organelle that extends from the surface of the cell. Found in both prokaryotes and eukaryotes.	Used for movement of the cell or to move fluids over the cell's surface for absorption.
Pilus (Pili pl.)	Hollow, hairlike structures made of protein found on prokaryote cells.	Allows cell to attach itself to another cell.
Ribosomes	Tiny, complex structures composed of protein and RNA. Often attached to endoplasmic reticulum.	Ribosomes are involved in protein synthesis.
Endoplasmic Reticulum	Extensive system of internal membranes.	Forms compartments to isolate cell substances.
Nucleus	Double membrane structure that encases chromosomes.	Cell structure which directs protein synthesis and cell reproduction.
Chromosomes	Long strands of DNA.	Store hereditary information.
Nucleolus	Aggregations of rRNA and ribosomal proteins.	Area were ribosomes are manufactured.
Golgi Complex	Flattened stacks of membranes.	Used in the collection, packaging, and distribution of synthesized molecules.
Peroxisomes	Membrane confined spherical body about 0.2 to 0.5 micrometers in diameter.	Formed by the endoplasmic reticulum. Converts fats into carbohydrates Detoxifies potentially harmful oxidants.
Lysosomes	Membrane confined spherical body about 0.2 to 0.5 micrometers in diameter.	Formed by the golgi complex. Contains digestive enzymes for braking down old cellular components.
Centrioles	Long hollow tubes composed of protein. Not found in plant cells.	Influence cell shape, move chromosomes during reproductive division, and are the internal structure for flagellum.
Secretory Vesicles	Membrane enclosed sack created at the golgi complex.	These structures contain cell secretions, like hormones and neurotransmitters. The secretory vesicles are then transported to the cell surface where they are release to the environment outside the cell.
Vacuole	Voids within the cytoplasm. Quite large in plant cells.	Used to store water and waste products.
Cytoplasm	Semifluid mixture that occupies most of the cell's interior. Contains sugars, amino acids, and proteins. Also, contains a protein fiber network.	Medium in which organelles and other internal structures exist in. Fiber network support the shape of the cell and anchor organelles to fixed positions.
Mitochondria	Elongated structures about 1 to 3 micrometers long. Resemble aerobic bacteria.	Structure which converts sugar into energy through oxidation.
Chloroplasts	Elongated structures with vesicles containing chlorophyll.	Site of photosynthesis.

Measurement of Energy

In the previous section, we developed the concept of energy. We now must be able to measure and quantify it, using a standard set of units. Worldwide, two systems of units of measurement are commonly used today: the Metric System (Systeme International) and the British System.

The units of energy described in these systems are derived from a technical definition of energy used by physicists. This definition suggests that energy can be represented by the following simple equation:

Work = Force × Distance

Similar to the definition given in the previous topic, physicists view energy as the ability to do work. However, they define work as a force applied to some form of matter (object) multiplied by the distance that this object travels. Physicists commonly describe force with a unit of measurement known as a newton (after Sir Isaac Newton). A newton is equal to the force needed to accelerate (move) a mass weighting one kilogram one meter in one second in a vacuum with no friction. The work or energy required to move an object with the force of one newton over a distance of one meter is called a joule.

Some other definitions for the energy measurement units that you may come across in this textbook are as follows:

Calorie - equals the amount of heat required to raise 1 gram of pure water from 14.5 to 15.5° Celsius at standard atmospheric pressure. 1 calorie is equal to 4.1855 joules. The abreviation for calorie is cal. A kilocalorie, abbreviated kcal, is equal to a 1000 calories. 1 kilocalorie is equal to 4185 joules.

Btu - also called British thermal unit is the amount of energy required to raise the temperature of one pound of water one degree Fahrenheit.

Watt (W/m^2 or Wm^{-2}) - a metric unit of measurement of the intensity of radiation in watts over a square meter surface. One watt is equal to one joule of work per second. A kilowatt (kW) is the same as 1000 watts.

ENERGY, TEMPERATURE, AND HEAT

So far we have learned that energy can take on many forms. One important form of energy, relative to life on Earth, is kinetic energy. Simply defined, kinetic energy is the energy of motion. The amount of kinetic energy that a body possesses is dependent on the speed of its motion and its mass. At the atomic scale, the kinetic energy of atoms and molecules is sometimes referred to as heat energy.

Kinetic energy is also related to the concept of temperature. Temperature is defined as the measure of the average speed of atoms and molecules. The higher the temperature, the faster these particles of matter move. At a temperature of -273.15° Celsius (absolute zero) all atomic motion stops.

Heat is often defined as energy in the process of being transferred from one object to another because of difference in temperature between them. Heat is commonly transferred around our planet by the processes of conduction, convection, advection, and radiation.

Some other important definitions related to energy, temperature, and heat are:

- *Heat Capacity* - is the amount of heat energy absorbed by a substance associated to its corresponding temperature increase.
- *Specific Heat* - is equivalent to the heat capacity of a unit mass of a substance or the heat needed to raise the temperature of one gram (g) of a substance one degree Celsius. Water requires about 4 to 5 times more heat energy to raise its temperature when compared to an equal mass of most types of solid matter. This explains why water bodies heat more slowly than adjacent land surfaces.
- *Sensible Heat* - is heat that we can sense. A thermometer can be used to measure this form of heat. Several different scales of measurement exist for measuring sensible heat. The most common are: Celsius scale, Fahrenheit scale, and the Kelvin scale.
- *Latent Heat* - is the energy needed to change a substance to a higher state of matter. This same energy is released from the substance when the change of state (or phase) is reversed. The diagram below describes the various exchanges of heat involved with 1 gram of water.

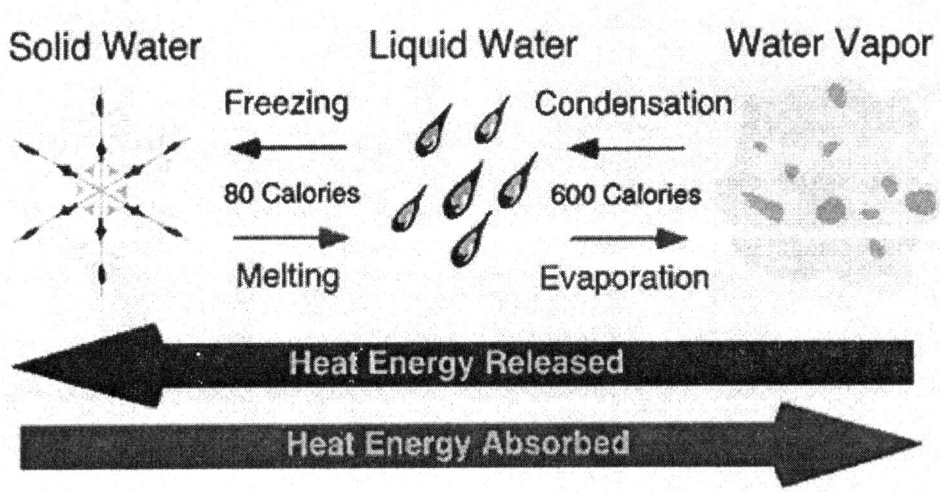

Figure: Latent heat exchanges of energy involved with the phase changes of water.

Figures below show the net absorption and release of latent heat energy for the Earth's surface for January and July, respectively. The highest values of flux or flow occur near the subtropical oceans where high temperatures and a plentiful supply of water encourage the evaporation of water. Negative values of latent heat flux indicate a net release of latent energy back into the environment because of the condensation or freezing of water. Values of latent heat flux are generally low over landmasses because of a limited supply of water at the ground surface.

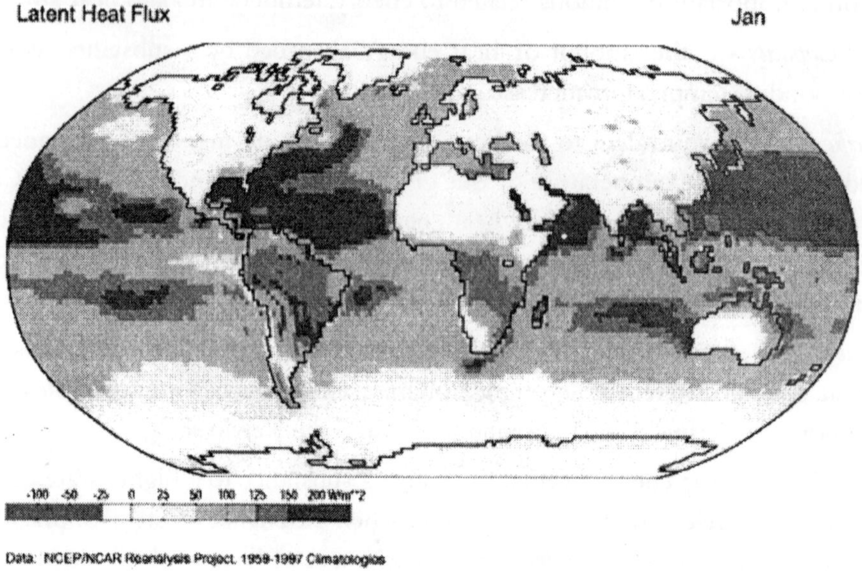

Figure: Mean January latent heat flux for the Earth's surface, 1959-1997.

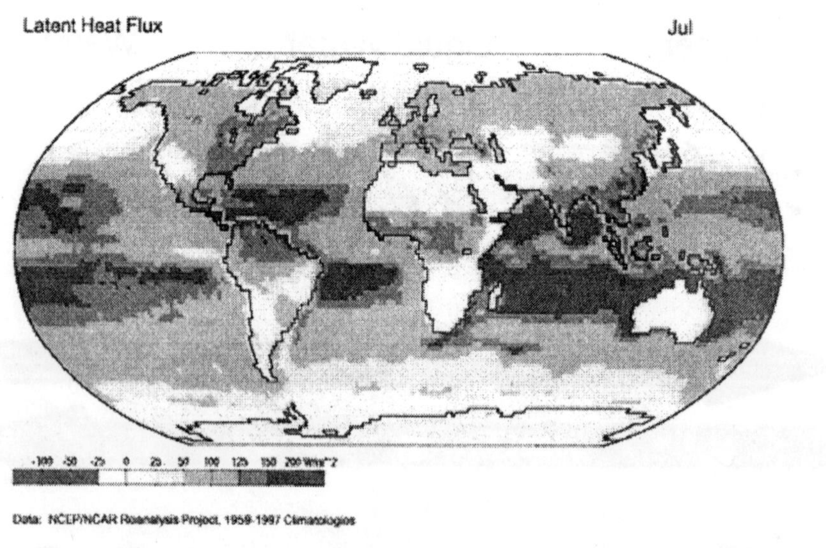

Figure: Mean July latent heat flux for the Earth's surface, 1959-1997.

Energy and Life

The capture and use of energy in living systems is dominated by two processes: photosynthesis and respiration. Through these two processes living organisms are able to capture and use all of the energy they require for their activities.

Photosynthesis

Plants can capture the electromagnetic energy from the Sun by a chemical process called photosynthesis. This chemical reaction can be described by the following simple equation:

$$6CO_2 + 6H_2O + \text{light energy} >>> C_6H_{12}O_6 + 6O_2$$

The product of photosynthesis is the carbohydrate glucose and oxygen which is released into the atmosphere. All of the sugar glucose is produced in the specialized photosynthetic cells of plants and some other organisms. Glucose is produced by chemically combining carbon dioxide and water with sunlight.

This chemical reaction is catalyzed by chlorophyll acting in concert with other pigment, lipid, sugars, protein, and nucleic acid molecules. Sugars created in photosynthesis can be later converted by the plant to starch for storage, or it can be combined with other sugar molecules to form specialized carbohydrates such as cellulose, or it can be combined with other nutrients such as nitrogen, phosphorus, and sulfur, to build complex molecules such as proteins and nucleic acids.

Because all the energy fixed by a plant is converted into sugar, it is theoretically possible to determine a plant's energy uptake by measuring the amount of sugar produced. This quantity is called gross primary productivity. Measurements of the buildup of sugar in the plant reflect gross primary productivity less respiration, or net primary productivity.

In general, animals cannot produce their own energy via photosynthesis. Instead, they capture their energy by the consumption and assimilation of the biomass of plants or other animals. Thus, animals get the energy they need for maintenance of their bodies tissues, growth, and reproduction indirectly from photosynthetic organisms.

Respiration

The oxidation of sugar by organisms is called respiration. This process occurs in both plants and animals. In most organisms, respiration releases the energy required for all metabolic processes. This chemical reaction can be described by the following simple equation:

$$C_6H_{12}O_6 + 6O_2 >>> 6CO_2 + 6H_2O + \text{released energy}$$

One of the products of respiration is energy, which is released via the chemical decomposition of glucose. Other products of this chemical reaction are carbon dioxide (CO_2) and water (H_2O).

Laws of Thermodynamics

The field of thermodynamics studies the behavior of energy flow in natural systems. From this study, a number of physical laws have been established. The laws of thermodynamics describe some of the fundamental truths of thermodynamics observed in our Universe. Understanding these laws is important to students of Physical Geography because many of the processes studied involve the flow of energy.

First Law of Thermodynamics

The first law of thermodynamics is often called the Law of Conservation of Energy. This law suggests that energy can be transferred from one system to another in many forms. Also, it can not be created or destroyed. Thus, the total amount of energy available in the Universe is constant. Einstein's famous equation (written below) describes the relationship between energy and matter:

$$E = mc^2$$

In the equation above, energy (E) is equal to matter (m) times the square of a constant (c). Einstein suggested that energy and matter are interchangeable. His equation also suggests that the quantity of energy and matter in the Universe is fixed.

Second Law of Thermodynamics

Heat cannot be transfer from a colder to a hotter body. As a result of this fact of thermodynamics, natural processes that involve energy transfer must have one direction, and all natural processes are irreversible.

This law also predicts that the entropy of an isolated system always increases with time. Entropy is the measure of the disorder or randomness of energy and matter in a system. Because of the second law of thermodynamics both energy and matter in the Universe are becoming less useful as time goes on. Perfect order in the Universe occurred the instance after the Big Bang when energy and matter and all of the forces of the Universe were unified.

Third Law of Thermodynamics

The third law of thermodynamics states that if all the thermal motion of molecules (kinetic energy) could be removed, a state called absolute zero would occur. Absolute zero results in a temperature of 0 Kelvins or -273.15° Celsius.

Absolute Zero = 0 Kelvins = -273.15° Celsius

The Universe will attain absolute zero when all energy and matter is randomly distributed across space. The current temperature of empty space in the Universe is about 2.7 Kelvins.

The Nature of Radiation

All objects above the temperature of absolute zero (-273.15° Celsius) radiate energy to their surrounding environment. This energy, or radiation, is emitted as electromagnetic waves that travel at the speed of light. Many different types of radiation have been identified. Each of these types is defined by its wavelength. The wavelength of electromagnetic radiation can vary from being infinitely short to infinitely long.

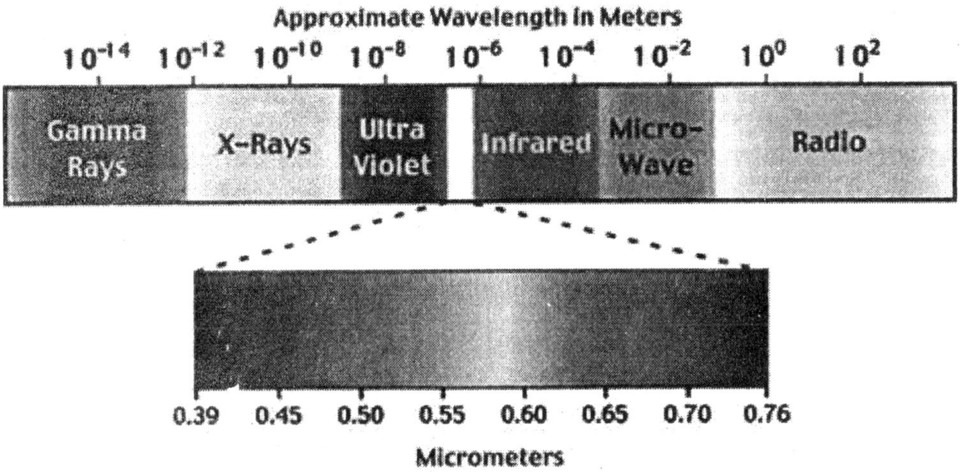

Figure: Some of the various types of electromagnetic radiation as defined by wavelength. Visible light has a spectrum that ranges from 0.40 to 0.71 micrometers (μm).

Visible light is a form of electromagnetic radiation that can be perceived by our eyes. Light has a wavelength of between 0.40 to 0.71 micrometers (μm). Figure above illustrates that various spectral color bands that make up light.

$$E^* = \sigma T^4$$

where σ (sigma) = $5.67 \times 10^{-8}\ Wm^{-2}\ K^{-4}$

and T is the temperature in Kelvin

According to the Stephan-Boltzmann equation, a small increase in the temperature of a radiating body results in a large amount of additional radiation being emitted.

The Sun emits only a portion (44 %) of its radiation in zone. Solar radiation spans a spectrum from approximately 0.1 to 4.0 micrometers. The band from 0.1 to 0.4 micrometers is called ultraviolet radiation. About 7% of the Sun's emission is in this wavelength band. About 48% of the Sun's radiation falls in the region between 0.71 to 4.0 micrometers. This band is called the near (0.71 to 1.5 micrometers) and far infrared (1.5 to 4.0 micrometers).

The amount of electromagnetic radiation emitted by a body is directly related to its temperature.

If the body is a perfect emitter (black body), the amount of radiation given off is proportional to the 4th power of its temperature as measured in Kelvin units. This natural phenomenon is described by the Stephan-Boltzmann Law. The following simple equation describes this law mathematically:

In general, good emitters of radiation are also good absorbers of radiation at specific wavelength bands.

This is especially true of gases and is responsible for the Earth's greenhouse effect. Likewise, weak emitters of radiation are also weak absorbers of radiation at specific wavelength bands. This fact is referred to as Kirchhoff's Law.

Some objects in nature have almost completely perfect abilities to absorb and emit radiation. We call these objects black bodies. The radiation characteristics of the Sun and the Earth are very close to being black bodies.

The wavelength of maximum emission of any body is inversely proportional to its absolute temperature. Thus, the higher the temperature, the shorter the wavelength of maximum emission. This phenomenon is often called Wien's Law. The following equation describes this law:

$$\lambda_{max} = C/T$$

where C is a constant equal to 2897
and T is the temperature in Kelvin

Wien's law suggests that as the temperature of a body increases, the wavelength of maximum emission becomes smaller.

According to the above equation the wavelength of maximum emission for the Sun (5800 Kelvins) is about 0.5 micrometers, while the wavelength of maximum emission for the Earth (288 Kelvins) is approximately 10.0 micrometers.

A graph that describes the quantity of radiation that is emitted from a body at particular wavelengths is commonly called a spectrum. The following two graphs describe the spectrums for the Sun and Earth.

Sun

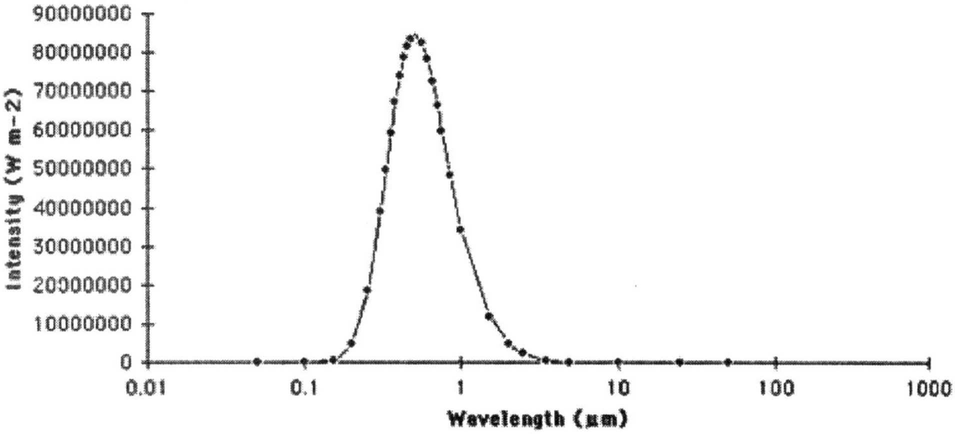

Figure: Spectrum of the Sun. The Sun emits most of its radiation in a wavelength band between 0.1 and 4.0 micrometers (µm).

Earth

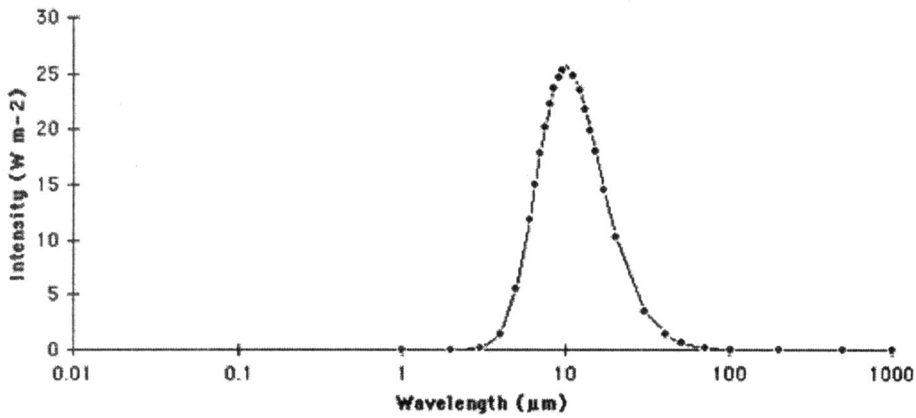

Figure: Spectrum of the Earth. The Earth emits most of its radiation in a wavelength band between 0.5 and 30.0 micrometers (µm).

The above graphs illustrate two important points concerning the relationship between the temperature of a body and its emissions of electromagnetic radiation:

1. The amount of radiation emitted from a body increases exponentially with a linear rise in temperature (see above Stephan-Boltzmann's Law).
2. The average wavelength of electromagnetic emissions becomes shorter with increasing temperature (see above Wien's Law).

Finally, the amount of radiation passing through a specific area is inversely proportional to the square of the distance of that area from the energy source. This phenomenon is called the Inverse Square Law. Using this law we can model the effect that distance traveled has on the intensity of emitted radiation from a body like the Sun. Figure below suggests that the intensity of radiation emitted by a body quickly diminishes with distance in a nonlinear fashion.

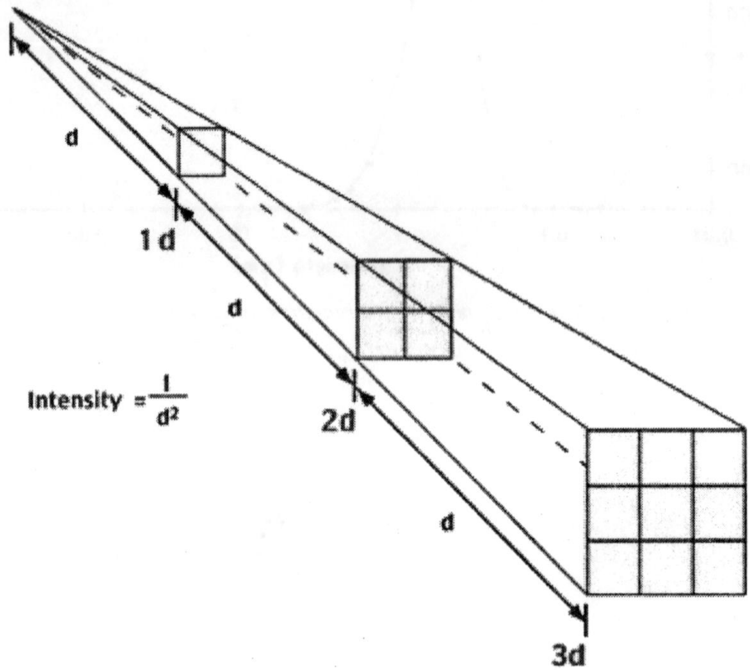

Figure: Diagram illustrating the diffusion of radiation due to the Inverse Square Law.

Mathematically, the Inverse Square Law is described by the equation:

Intensity = I/d^2

where I is the intensity of the radiation at 1d (see above diagram) and d is the distance traveled.

Let us try this equation out. For example, what would be the intensity of emitted radiation traveling two units of distance if the intensity at 1d = 90,000 units?

Intensity = 90,000/22

Intensity = 90,000/4

Intensity = 22,500 units

What would be the intensity of emitted radiation traveling three units of distance if the intensity at 1d = 90,000 units?

Intensity = 90,000/32

Intensity = 90,000/9

Intensity = 10,000 units

What would be the intensity of emitted radiation traveling four units of distance if the intensity at 1d = 90,000 units?

Intensity = 90,000/42

Intensity = 90,000/16

Intensity = 5625 units

Note that the decrease in intensity with distance is not linear when graphed!

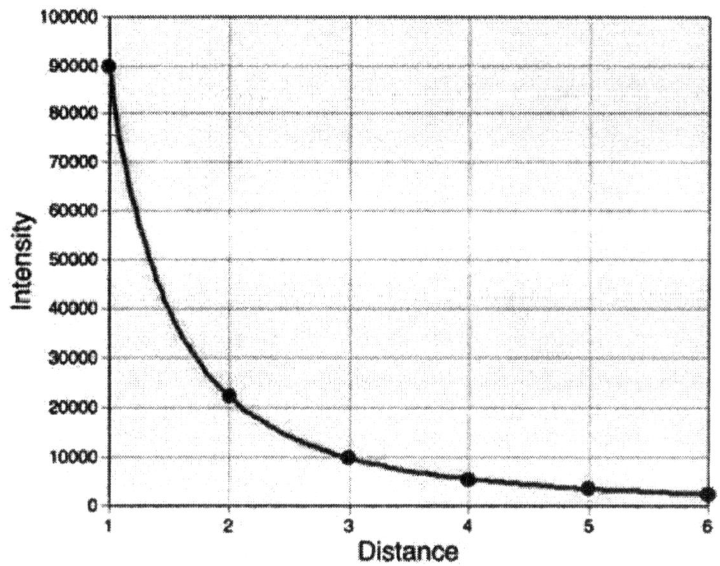

Figure: Reduction in intensity of radiation with distance traveled.

The Solar Source of the Earth's Energy

Almost all of the energy that drives the various systems (climate systems, ecosystems, hydrologic systems, etc.) found on the Earth originates from the Sun.

Solar energy is created at the core of the Sun when hydrogen atoms are fused into helium by nuclear fusion.

For each second of this nuclear process, 700 million tons of hydrogen are converted into 695 million tons of helium. The remaining 5 million tons are turned into electromagnetic energy that radiates from the Sun's surface out into space.

The radiative surface of the Sun, or photosphere, has an average temperature of about 5800 Kelvins. Most of the electromagnetic radiation emitted from the Sun's surface lies in the visible band centered at 0.5 μm. The total quantity of energy emitted from the Sun's surface is approximately 63,000,000 Watts per square meter (W/m^2 or Wm^{-2}).

The energy emitted by the Sun passes through space until it is intercepted by planets and other celestial objects. The intensity of solar radiation striking these objects is determined by a physical law known as the Inverse Square Law.

This law merely states that the intensity of the radiation emitted from the Sun varies with the squared distance from the source. As a result of this law, if the intensity of radiation at a given distance is one unit, at twice the distance the intensity will become only one-quarter. At three times the distance, the intensity will become only one-ninth of its original intensity at a distance of one unit, and so on.

Given the amount of energy radiated by the Sun and the average Earth-Sun distance of 149.5 million kilometers, the amount of radiation intercepted by the outer limits of the atmosphere can be calculated to be around 1370 W/m^2. For general purposes, the energy output of the Sun can be considered constant.

This of course is not entirely true. Scientists have shown that the output of the Sun is temporally variable. Some researchers have also suggested that the increase in the average global temperature over the last century may have been solar in origin. This statement, however, is difficult to prove because accurate data on solar output of radiation only goes back to about 1978.

EARTH-SUN GEOMETRY

Earth Rotation and Revolution

The term Earth rotation refers to the spinning of our planet on its axis. Because of rotation, the Earth's surface moves at the equator at a speed of about 467 m per second or slightly over 1675 km per hour.

If you could look down at the Earth's North Pole from space you would notice that the direction of rotation is counter-clockwise.

The opposite is true if the Earth is viewed from the South Pole. One rotation takes exactly twenty-four hours and is called a mean solar day. The Earth's rotation is responsible for the daily cycles of day and night.

At any one moment in time, one half of the Earth is in sunlight, while the other half is in darkness. The edge dividing the daylight from night is called the circle of illumination. The Earth's rotation also creates the apparent movement of the Sun across the horizon.

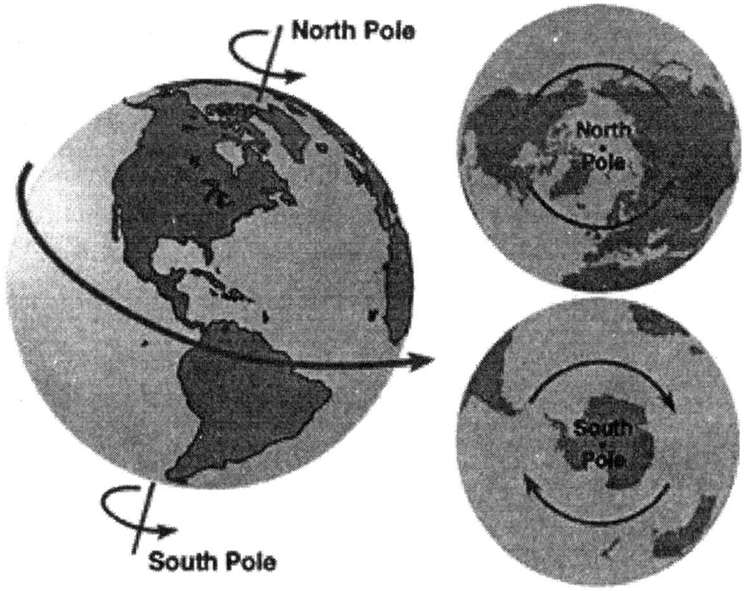

Figure: The movement of the Earth about its axis is known as rotation. The direction of this movement varies with the viewer's position. From the North Pole the rotation appears to move in a counter-clockwise fashion. Looking down at the South Pole the Earth's rotation appears clockwise.

The orbit of the Earth around the Sun is called an Earth revolution. This celestial motion takes 365.26 days to complete one cycle. Further, the Earth's orbit around the Sun is not circular, but oval or elliptical. An elliptical orbit causes the Earth's distance from the Sun to vary over a year. Yet, this phenomenon is not responsible for the Earth's seasons!

This variation in the distance from the Sun causes the amount of solar radiation received by the Earth to annually vary by about 6%. Figure below illustrates the positions in the Earth's revolution where it is closest and farthest from the Sun.

On January 3, perihelion, the Earth is closest to the Sun (147.3 million km). The Earth is farthest from the Sun on July 4, or aphelion (152.1 million km). The average distance of the Earth from the Sun over a one-year period is about 149.6 million km.

TILT OF THE EARTH'S AXIS

The ecliptic plane can be defined as a two-dimensional flat surface that geometrically intersects the Earth's orbital path around the Sun. On this plane, the Earth's axis is not at right angles to this surface, but inclined at an angle of about 23.5° from the perpendicular. Figure below shows a side view of the Earth in its orbit about the Sun on four important dates: June solstice, September equinox, December solstice, and March equinox. Note that the angle of the Earth's axis in relation to the ecliptic plane and the North Star on these four dates remains unchanged. Yet, the relative position of the Earth's axis to the Sun does change during this cycle. This circumstance is responsible for the annual changes in the height of the Sun above the horizon. It also causes the seasons, by controlling the intensity and duration of sunlight received by locations on the Earth. Figure below shows an overhead view of this same phenomenon. In this view, we can see how the circle of illumination changes its position on the Earth's surface. During the two equinoxes, the circle of illumination cuts through the North Pole and the South Pole. On the June solstice, the circle of illumination is tangent to the Arctic Circle (66.5° N) and the region above this latitude receives 24 hours of daylight. The Arctic Circle is in 24 hours of darkness during the December solstice.

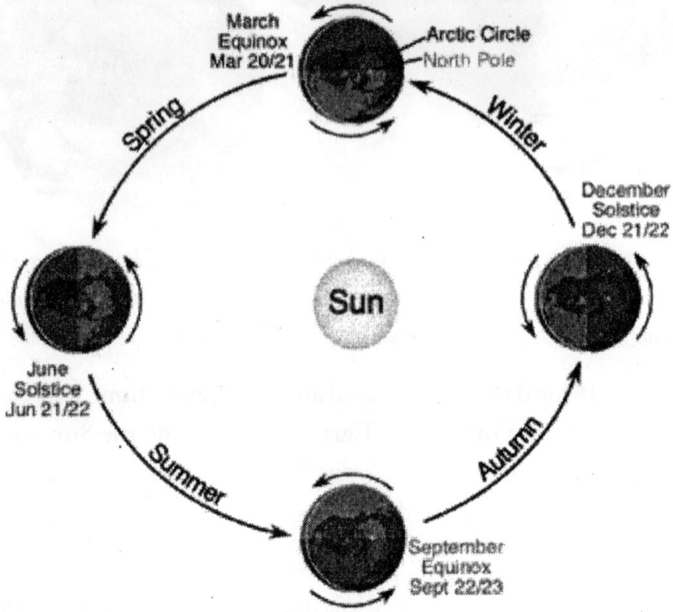

Figure: Annual change in the position of the Earth in its revolution around the Sun. In this graphic, we are viewing the Earth from a position in space that is above the North Pole (yellow dot) at the summer solstice, the winter solstice, and the two equinoxes. Note how the position of the North Pole on the Earth's surface does not change. However, its position relative to the Sun does change and this shift is responsible for the seasons. The red circle on each of the Earths represents the Arctic Circle (66.5 degrees N). During the June solstice, the area above the Arctic Circle is experiencing 24 hours of daylight because the North Pole is tilted 23.5 degrees toward the Sun. The Arctic Circle experiences 24 hours of night when the North Pole is tilted 23.5 degrees away from the Sun in the December solstice. During the two equinoxes, the circle of illumination cuts through the polar axis and all locations on the Earth experience 12 hours of day and night. Seasons are appropriate only for the Northern Hemisphere.

On June 21 or 22 the Earth is positioned in its orbit so that the North Pole is leaning 23.5° toward the Sun. During the June solstice (also called the summer solstice in the Northern Hemisphere), all locations north of the equator have day lengths greater than twelve hours, while all locations south of the equator have day lengths less than twelve hours. On December 21 or 22 the Earth is positioned so that the South Pole is leaning 23.5 degrees toward the Sun. During the December solstice (also called the winter solstice in the Northern Hemisphere), all locations north of the equator have day lengths less than twelve hours, while all locations south of the equator have day lengths exceeding twelve hours.

On September 22 or 23, also called the autumnal equinox in the Northern Hemisphere, neither pole is tilted toward or away from the Sun. In the Northern Hemisphere, March 20 or 21 marks the arrival of the vernal equinox or spring when once again the poles are not tilted toward or away from the Sun. Day lengths on both of these days, regardless of latitude, are exactly 12 hours.

Axis Tilt and Solar Altitude

The annual change in the relative position of the Earth's axis in relationship to the Sun causes the height of the Sun or solar altitude to vary in our skies.

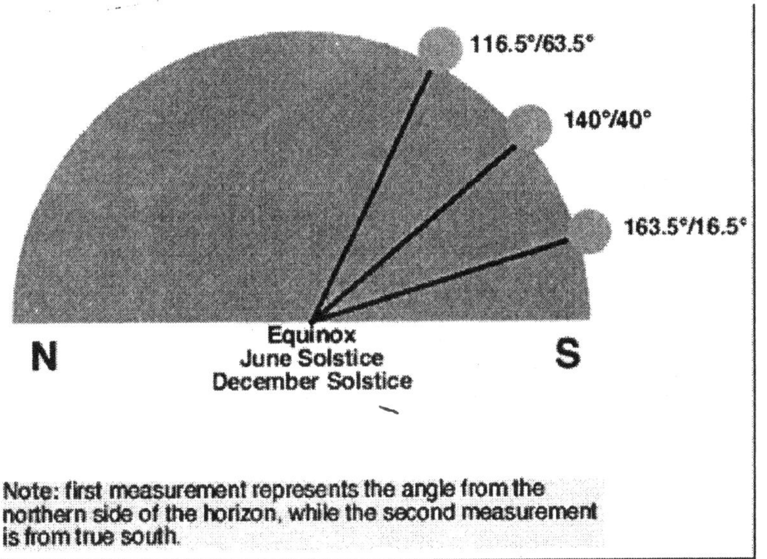

Figure: Variations in solar altitude at solar noon for 50 degrees North during the June solstice, equinox, and December solstice.

Solar altitude is normally measured from either the southern or northern point along the horizon and begins at zero degrees. Maximum solar altitude occurs when the Sun is directly overhead and has a value of 90°. The total variation in maximum solar altitude for any location on the Earth over a one-year period is 47° (Earth's tilt 23.5° × 2 = 47°). This variation is due to the annual changes in the relative position of the Earth to the Sun. At 50 degrees North, maximum solar altitude varies from 63.5 degrees on the June solstice to 16.5 degrees on the December solstice.

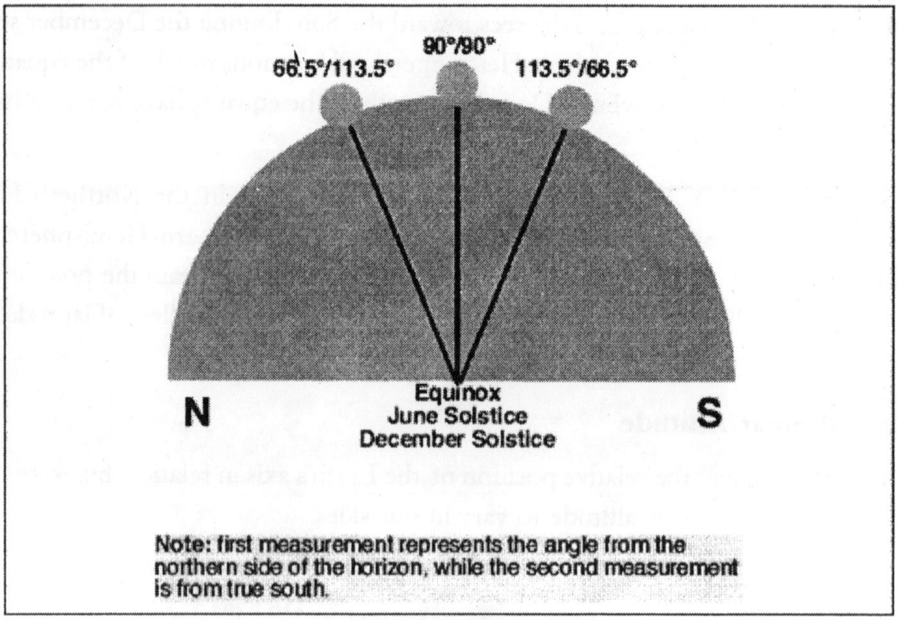

Figure: Variations in solar altitude at solar noon for the equator during the June solstice, equinox, and December solstice.

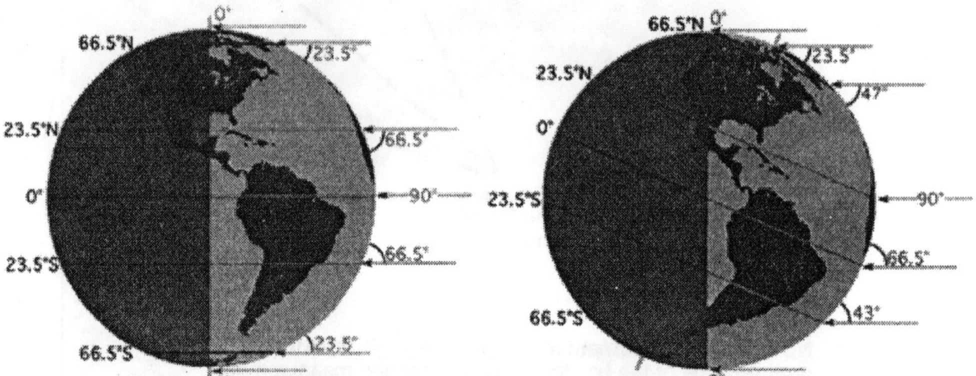

Figure: Relationship of maximum Sun height to latitude for the equinox (left) and June solstice (right). The red values on the right of the globes are maximum solar altitudes at solar noon. Black numbers on the left indicate the location of the Equator, Tropic of Cancer (23.5 degrees N), Tropic of Capricorn (23.5 degrees S), Arctic Circle (66.5 degrees N), and the Antarctic Circle (66.5 degrees S). The location of the North and South Poles are also identified. During the equinox, the equator is the location on the Earth with a Sun angle of 90 degrees for solar noon. Note how maximum Sun height declines with latitude as you move away from the Equator. For each degree of latitude traveled maximum Sun height decreases by the same amount. At equinox, you can also calculate the noon angle by subtracting the location's latitude from 90. During the summer solstice, the Sun is now directly overhead at the Tropic of Cancer. All locations above this location have maximum Sun heights that are 23.5 degrees higher from the equinox situation. Places above the Arctic Circle are in 24 hours of daylight. Below the Tropic of Cancer the noon angle of the Sun drops one degree in height for each degree of latitude traveled. At the Antarctic Circle, maximum Sun height becomes 0 degrees and locations south of this point on the Earth are in 24 hours of darkness.

Maximum solar height at the equator goes from 66.5 degrees above the northern end of the horizon during the June solstice, to directly overhead on the September equinox, and then down to 66.5 degrees above the southern end of the horizon during the December solstice.

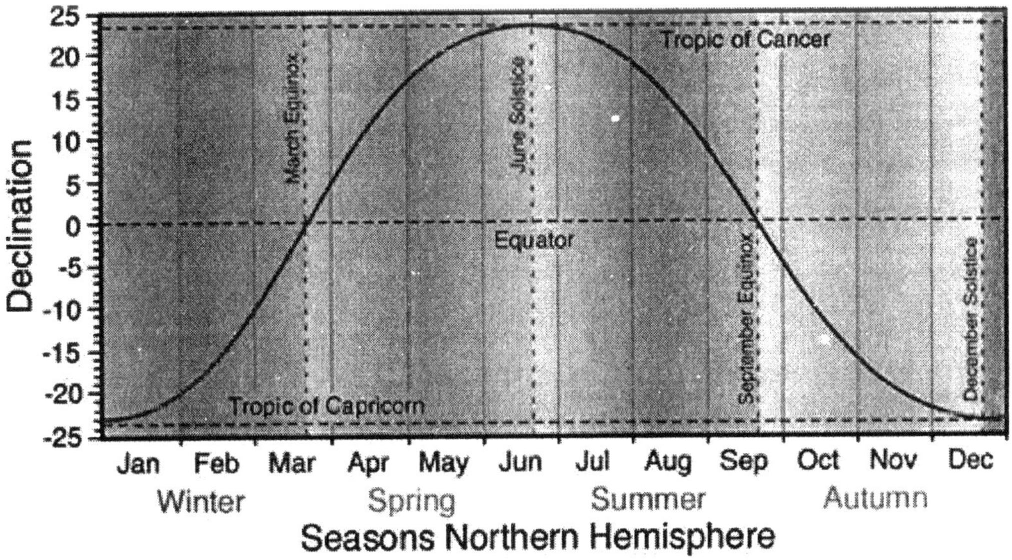

Figure: Angle of the Sun's declination and latitude of the subsolar point throughout the year. Seasons are for the Northern Hemisphere.

The location on the Earth where the Sun is directly overhead at solar noon is known as the subsolar point.

The subsolar point occurs on the equator during the two equinoxes. On these dates, the equator is lined up with the ecliptic plane and the poles are in line with the circle of illumination.

During the summer solstice, the subsolar point moves to the Tropic of Cancer (23.5° N) because at this time the North Pole is inclined 23.5° toward the Sun.

Figure below shows how the subsolar point gradually changes from one day to the next over a period of one-year. Note that on this graph, the subsolar point is located at the Tropic of Capricorn (23.5° S) during the December solstice when the South Pole is angled 23.5° toward the Sun.

The following table describes the changes in solar altitude at solar noon for the two solstices and equinoxes.

All measurements are in degrees (horizon has 180 degrees from True North to True South) and are measured from either True North or True South (whatever is closer).

Table: Maximum Sun altitudes for selected latitudes during the two solstices and equinoxes.

Location's Latitude	March Equinox March 20/21	June Solstice June 21/22	September Equinox September 22/23	December Solstice December 21/22
90 N	0 degrees	23.5 degrees	0 degrees	- 23.5 degrees
70 N	20 degrees	43.5 degrees	20 degrees	-3.5 degrees
66.5 N	23.5 degrees	47 degrees	23.5 degrees	0 degrees
60 N	30 degrees	53.5 degrees	30 degrees	6.5 degrees
50 N	40 degrees	63.5 degrees	40 degrees	16.5 degrees
23.5 N	66.5 degrees	90 degrees	66.5 degrees	43 degrees
0 degrees	90 degrees	66.5 degrees	90 degrees	66.5 degrees
23.5 S	66.5 degrees	43 degrees	66.5 degrees	90 degrees
50 S	40 degrees	16.5 degrees	40 degrees	63.5 degrees
60 S	30 degrees	6.5 degrees	30 degrees	53.5 degrees
66.5 S	23.5 degrees	0 degrees	23.5 degrees	47 degrees
70 S	20 degrees	-3.5 degrees	20 degrees	43.5 degrees
90 S	0 degrees	- 23.5 degrees	0 degrees	23.5 degrees

The following links show graphical illustrations of the annual movements of the Sun in our skies for selected latitudes. In these illustrations, solar angles are measured from both True North and True South for solar noon.

| 90 N | 66.5 N | 50 N | 23.5 N | Equator (0) | 23.5 S | 50 S | 66.5 S | 90 S |

Finally, the altitude of the Sun at solar noon can also be calculated with the following simple equation:

Altitude A = 90 - Latitude L +/- Declination D

In this equation, L is the latitude of the location in degrees and D is the declination. The equation is simplified to A = 90 - L if Sun angle determinations are being made for either equinox date. If the Sun angle determination is for a solstice date, declination (D) is added to latitude (L) if the location is experiencing summer (northern latitudes = June solstice; southern latitudes = December solstice) and subtracted from latitude (L) if the location is experiencing winter (northern latitudes = December solstice; southern latitudes = June solstice). All answers from this equation are given relative to True North for southern latitudes and True South for northern latitudes. For our purposes only the declinations of the two solstices and two equinoxes are important. These values are: June solstice D=23.5, December solstice D=23.5, March equinox D=0, and Septemeber equinox D=0.

When using the above equation in tropical latitudes, Sun altitude values greater than 90 degrees may occur for some calculations. When this occurs, the noonday Sun is actually behind you when looking towards the equator. Under these circumstances, Sun altitude should be recalculated as follows:

Altitude A = 90 - (originally calculated Altitude A - 90)

Earth-Sun Relationships and Insolation

In the previous topic, we learned that the Earth's seasons are controlled by changes in the duration and intensity of solar radiation or insolation. Both of these factors are in turn governed by the annual change in the position of the Earth's axis relative to the Sun.

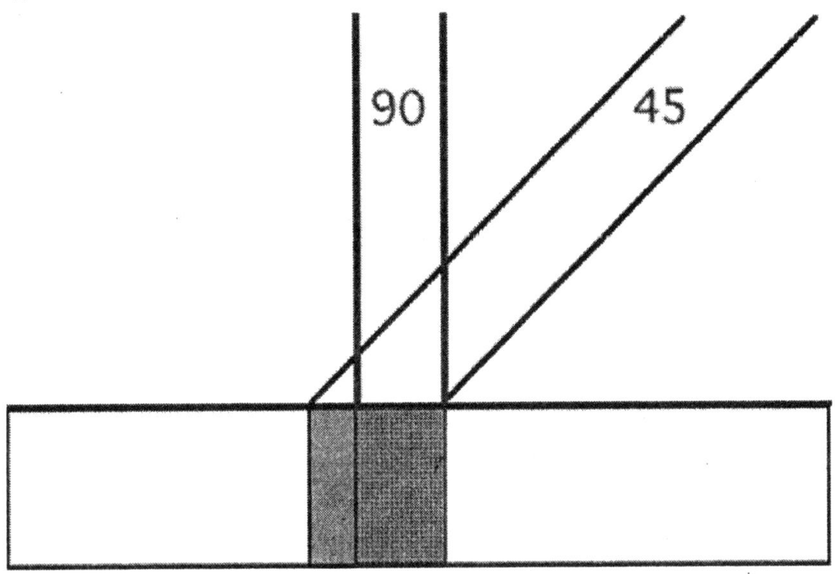

Figure: Effect of angle on the area that intercepts an incoming beam of radiation.

Yearly changes in the position of the Earth's axis cause the location of the Sun to wander 47° across our skies. Changes in the location of the Sun have a direct effect on the intensity of solar radiation.

The intensity of solar radiation is largely a function of the angle of incidence, the angle at which the Sun's rays strike the Earth's surface. If the Sun is positioned directly overhead or 90° from the horizon, the incoming insolation strikes the surface of the Earth at right angles and is most intense. If the Sun is 45° above the horizon, the incoming insolation strikes the Earth's surface at an angle. This causes the rays to be spread out over a larger surface area reducing the intensity of the radiation.

Figure below models the effect of changing the angle of incidence from 90 to 45°. As illustrated, the lower Sun angle (45°) causes the radiation to be received over a much larger surface area. This surface area is approximately 40% greater than the area covered by an angle of 90°. The lower angle also reduces the intensity of the incoming rays by 30%.

We can also model the effect the angle of incidence has on insolation intensity with the following simple equation:

Intensity = SIN (A)

where, A is the angle of incidence and SIN is the sine function found on most calculators. Using this equation we can determine that an angle of 90° gives us a value of 1.00 or 100% (1.00 x 100). Let us compare this maximum value with values determined for other angles of incidence. Note the answers are expressed as a percentage of the potential maximum value.

SIN 80 = 0.98 or 98%

SIN 70 = 0.94 or 94%

SIN 60 = 0.87 or 87%

SIN 50 = 0.77 or 77%

SIN 40 = 0.64 or 64%

SIN 30 = 0.50 or 50%

SIN 20 = 0.34 or 34%

SIN 10 = 0.17 or 17%

SIN 0 = 0.00 or 0%

The yearly changes in the position of the Earth's axis relative to the plane of the ecliptic also causes seasonal variations in day length to all locations outside of the equator. Longest days occur during the June solstice for locations north of the equator and on the December solstice for locations in the Southern Hemisphere.

The equator experiences equal day and night on every day of the year. Day and night is also of equal length for all Earth locations on the September and March equinoxes. Figure below describes the change in the length of day for locations at the equator, 10, 30, 50, 60, and 70 degrees North over a one-year period. The illustration suggests that days are longer than nights in the Northern Hemisphere from the March equinox to the September equinox. Between the September to March equinox days are shorter than nights in the Northern Hemisphere. The opposite is true in the Southern Hemisphere. The graph also shows that the seasonal (winter to summer) variation in day length increases with increasing latitude.

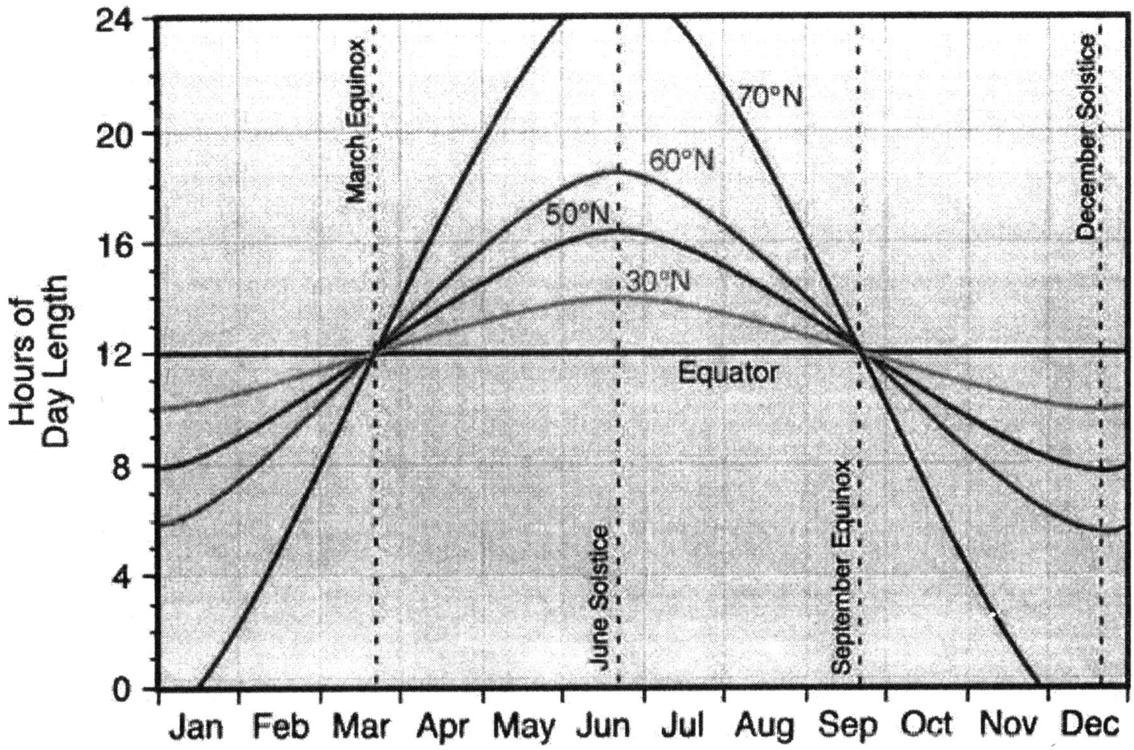

Figure: Annual variations in day length for locations at the equator, 30, 50, 60, and 70° North latitude.

Figure below describes the potential insolation available for the equator and several locations in the Northern Hemisphere over a one-year period. The values plotted on this graph take into account the combined effects of angle of incidence and day length duration. Locations at the equator show the least amount of variation in insolation over a one-year period.

These slight changes in insolation result only from the annual changes in the altitude of the Sun above the horizon, as the duration of daylight at the equator is always 12 hours. The peaks in insolation intensity correspond to the two equinoxes when the Sun is directly overhead. The two annual minimums of insolation occur on the solstices when the maximum height of the Sun above the horizon reaches an angle of 66.5°.

The most extreme variations in insolation received in the Northern Hemisphere occur at 90 degrees North. During the June solstice this location receives more potential incoming solar radiation than any other location graphed. At this time the Sun never sets. In fact, it remains at an altitude of 23.5 degrees above the horizon for the whole day. From September 22 (September equinox) to March 21, (March equinox) no insolation is received at 90 degrees North. During this period the Sun slips below the horizon as the northern axis of the Earth has an orientation that is tilted away from the Sun.

The annual insolation curve for locations at 60 degrees North best approximates the seasonal changes in solar radiation intensity perceived at our latitude. Maximum values of insolation are received at the June solstice when day length and angle of incidence are at their maximum. During the June solstice day length is 18 hours and 27 minutes and the angle of the Sun reaches a maximum value of 53.5 degrees above the horizon. Minimum values of insolation are received during the December solstice when day length and angle of incidence are at their minimum. During the December solstice day length is only 5 hours and 33 minutes and the angle of the Sun reaches a lowest value of 6.5 degrees above the horizon.

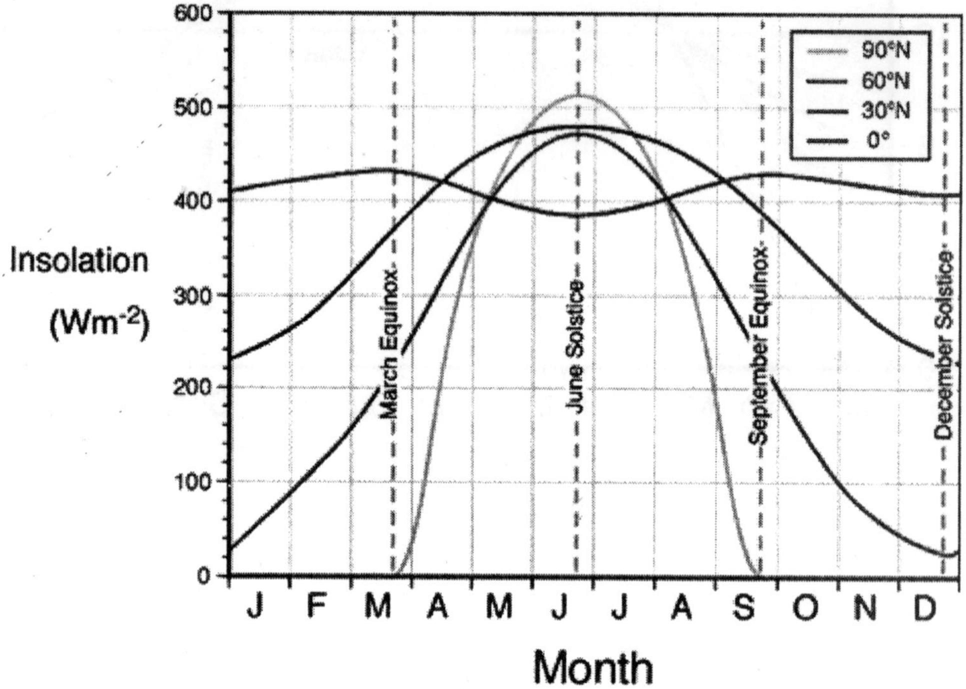

Figure: Monthly values of available insolation in Wm-2 for the equator, 30, 60, and 90° North.

BIBLIOGRAPHY

Agnew, John, David N. Livingstone, and Alisdair Rogers: *Human Geography: An Essential Anthology.* Oxford and Cambridge, Mass.: Blackwell, 1996.

Bernauer, Thomas: *Genes, Trade, and Regulation: The Seeds of Conflict in Food Biotechnology*, Princeton, Princeton University. 2003.

Clark, Michael J., Kenneth J. Gregory, and Angela M. Gurnell: *Horizons in Physical Geography.* Basingstoke and London: Macmillan Education, 1987.

Cosgrove, Denis: *Apollo's Eye: A Cartographic Genealogy of the Earth in the Western Imagination.* Baltimore and London: John Hopkins University Press, 2001.

Costanza, R.: *Ecological Economics*, Columbia Uni. Press., New York, 1991.

Diamond, Jared: *Guns, Germs, and Steel: The Fates of Human Societies.* New York: Norton, 1999.

Daubenmire, F.: *Plants and Environment*, New York, Wiley, 1947.

E. Dereniak and D. Crowe, *Optical Radiation Detectord*, John Wiley and Sons, 1984.

Edward A.: *Papers on Bacterial Genetics*, Boston, Brown and Company, 1960.

Entrikin, J. Nicholas, and Stanley D. Brunn: *Reflections on Richard Hartshorne's "The Nature of Geography."* Washington, D.C.: Association of American Geographers, 1989.

Haggett, Peter: *Geography: A Modern Synthesis.* 2nd ed. New York: Harper and Row, 1975.

Hartshorne, Richard: *Perspective on the Nature of Geography.* Chicago: Published by Rand McNally and Company as Monograph No. 1, 1959.

Harvey, David: *Spaces of the Capital: Towards a Critical Geography.* Edinburgh: Edinburgh University Press, 2001.

Katz, Cindi, and Janis Monk: *Full Circles: Geographies of Women over the Life Course.* London and New York: Routledge, 1993.

Lewis, Martin W., and Karen E. Wilgen: *The Myth of Continents: A Critique of Metageography.* Berkeley and Los Angeles: University of California Press, 1997.

Livingston, David N., and Charles W. J. Withers: *Geography and Enlightenment.* Chicago: University of Chicago Press, 1999.

Lodish, H.: *Molecular Cell Biology*, New York, Freeman, 2000.

Quaino, Massimo: *Geography and Marxism.* Translated by Alan Braley. Totowa, N.J.: Barnes and Noble, 1982.

Romm, James S.: *The Edges of the Earth in Ancient Thought: Geography, Exploration, and Fiction.* Princeton, N.J.: Princeton University Press, 1992.

Sheldon, J.K.: *Practical Environmental Bioremediation*, Lewis Publishers, Boca Raton, 1992.

Slater, R.J.: *Experiments in Molecular Biology*, Clifton, Humana Press, 1986.

Thrower, Norman J. W.: *Maps and Civilization: Cartography in Culture and Society.* Chicago and London: University of Chicago Press, 1999.

Teich, M.: *Fundamentals of Photonics*, New York: Wiley, 1991.

Walden, Richard: *Genetic Transformation in Plants*, England, Open University Press, 1988.

Wood, E.J. Ferguson: *Marine Microbial Ecology*, London, Chapman and Hall, 1965.

INDEX

□□□